Private Sector Involvement in Urban Solid Waste Collection

Sampson Oduro-Kwarteng

CRC Press/Belkema is an imprint of the Taylor & Francis Group, an informa business

© 2011, Sampson Oduro-Kwarteng

Published by:
CRC Press /Balkema
PO Box 447, 2300 AK Leiden, The Netherlands
e-mail: Pub.NL@tandf.co.uk
www.crcpress.com, www.balkema.nl, www.taylorandfrancis.co.uk,

ISBN 978-0-415-69277-9 (Taylor & Francis Group)

Private Sector Involvement in Urban Solid Waste Collection

Performance, Capacity, and Regulation in Five Cities in Ghana

De rol van de private sector bij het ophalen van vast afval in de stad. Prestaties, capaciteit en regulering in vijf steden in Ghana

Thesis

to obtain the Degree of Doctor from the
Erasmus University, Rotterdam
by command of the rector magnificus

Professor dr H.G. Schmidt

and in accordance with the decision of the Doctorate Board

The public defence shall be held on
Monday, 29 August 2011 at 16.00 hrs,
at ISS in The Hague, The Netherlands

by

Sampson Oduro-Kwarteng
Born in Dunkwa, Ghana

iSS International
Institute of Social Studies

Doctoral Committee

Promotor
Prof. dr. M.P. van Dijk

Other members
Prof. dr. L. van den Berg
Prof. dr. A.H.J. Helmsing
Prof. dr. M.A.R.M. Salih

Dedication

I dedicate this thesis to my wife Veronica and our lovely children, Priscilla, Daniel Caleb and Leslie Derrick, to inspire them to greater heights with the help of the Almighty God.

Contents

List of Tables

List of Figures

List of Boxes

Abbreviations

AMA	Accra Metropolitan Assembly
CBO	Community Based Organizations
CCC	Central Container Collection (communal collection)
EHD	Environmental Health Department
EHO	Environmental Health Officer
EHT	Environmental Health Technology
EPA	Environmental Protection Agency
ESPA	Environmental Service Providers Association
HtH	House-to-house Collection
KMA	Kumasi Metropolitan Assembly
KMA	Kumasi Metropolitan Assembly
LG	Local Government
LGs	Local Governments
MA	Metropolitan/Municipal Assembly
MDG	Millennium Development Goal
MLGRD	Ministry of Local Government and Rural Development
MMDAs	Metropolitan, Municipal, and District Assemblies
MSW	Municipal Solid Waste
NGO	Non Governmental Organisation
NPM	New Public Management
PPP	Public-Private Partisanship
PSP	Public Sector Participation
PSI	Private Sector Involvement
PAYT	Pay-As-You-Throw
STMA	Sekondi Takoradi Metropolitan Assembly
SWM	Solid Waste Management
SWC	Solid Waste Collection
TaMA	Tamale Municipal Assembly
TMA	Tema Municipal Assembly
UESP	Urban Environmental Sanitation Project
WMD	Waste Management Department

Acknowledgment

To Almighty God be the glory for the strength and great things he has done for me in pursuing this research study. This study would not have come to completion without enormous support and constructive comments from my supervisor. I am very thankful to my supervisor Professor Meine Pieter van Dijk for his invaluable support, time, guidance, and constructive comments.

I am grateful to Dr Nicolas Awortwi of ISS The Hague, Dr Klaas Sckwartz of UNESCO IHE, and Dr. K.B. Nyarko of KNUST for their useful comments. This thesis has come to completion not only by my own effort but also my students who assisted in the field data collection. I am grateful to all my students who made this research in five cities possible, especially Isaac Nooni, Edem Yeboah, M. Osei Bonsu, Frank Owusu Ansah, Yaayin Boniface, Prince Osei Bonsu, Barbara Korsah, and Saeed Munir.

I am also grateful to my colleagues at the Department of Civil Engineering, KNUST, for their encouragement especially, Prof. S. N. Odai, Prof. Esi Awuah, Dr. F.O.K. Anyemedu, Dr. G. Anornu, Dr. Richard Buamah, Mrs. Helen Essandoh, Mr. Frank Annor, Mr. Kwaku Amaning Adjei, Mr Charles Adams and Dr Adom Asamoah. They deserve special mention for their support and encouragement. I also appreciate the encouragement of UNESCO-IHE PhD students that I had the opportunity of interacting with in the course of the programme particularly, E. Owusu Ansa and Eric Ofosu and others who shared experiences.

I appreciate the invaluable support and cooperation from the Directors and Staff of Waste Management Departments of Accra, Tema, Kumasi, Takoradi and Tamale Metropolitan Assemblies, especially Mr. Anthony Mensah, Mr Prosper Kotoka, Mr Atteh Donkor, Mr E. Mba, Mr. Glory of STMA, Mr S. Kpodo. I am also thankful to all the directors and staff of the waste companies in the five cities, especially Yafuru Waste, Almanuel Waste, Liberty Waste and others for their support for providing data and responding to interviews and questionnaires.

I am thankful to all my friends, Mr Boakye in Delft, Miss Mizpah Asese, Dr. Patrick Sarfo, Mr E. Johnson and others for their support, prayers, and urging me on. To my siblings, in-laws and those I may have forgotten to mention I say thank you.

Finally, I am indebted to my loving wife Veronica Oduro-Kwarteng for her support, encouragement and prayers and to our wonderful kids, Priscilla Oduro-Kwarteng, Daniel Caleb Oduro-Kwarteng, and Leslie Derrick Oduro-Kwarteng for their prayers and feeling home sick in my absence.

Private Sector Involvement in Urban Solid Waste Collection

Performance, Capacity, and Regulation in Five Cities in Ghana

Abstract

This thesis focuses on the private sector involvement in solid waste collection, and the influence of private sector capacity and local governments' regulations on private sector performance. Private sector involvement in public service provision evolved to deal with market and government failures (van Dijk, 2008b). The public sector is playing a leading role of purchaser (buyer) on behalf of citizens through subsidies and/or user charges, whiles the private sector is taking on an increasing role as provider (seller) of public services, and being regulated by the public sector to correct market failures associated with 'public good' nature of solid waste service. It is believed that private sector involvement is a way to maintain market discipline and to bring private sector management and technical expertise and private finance into public service to achieve cost efficiency and better service provision (improve service quality) (Cointreau, 1994; Bartone, 2001; van Dijk, 2008b). Even though private sector involvement (PSI) provides opportunities for improved efficiency and service quality, the few studies on PSI in solid waste management in developing countries did not explore the private sector performance to ascertain the expected gains and the factors influencing efficiency and service quality at the firm level.

The main objective of this research is to examine the evolving involvement of private sector in urban solid waste collection, and the factors explaining differences in performance of private sector companies. The main research question is: To what extent do capacity and regulation for solid waste provision influence private sector performance? In order to establish the magnitude and causes of performance problems and possible policy interventions to these problems, the following questions are addressed in this study: How has private sector involvement evolved? What is the level of company performance (productivity and service quality)? What internal and external factors explain the level of performance?

The study was conducted in the five largest cities (Accra, Kumasi, Tema, Takoradi, and Tamale) in Ghana. These cities were chosen for the study because of their rapid urbanization and expansion. Qualitative and quantitative data were collected from five public organizations and twenty-five private companies. Secondary data on companies in the five cities were used for assessment of company productivity. A random household survey of 1500 households was conducted in the five cites. Households' perceptions about the existing service quality were rated on a scale from one to five and scores calculated. Household data were analyzed to determine the quality of service and residents' satisfaction. The implications of the study for performance improvement, regulatory policy, and sustainable service delivery have been explored.

Results on private sector performance show that there was a significant difference in service quality across cities, and that the service quality of franchised service (house-to-house collection) was higher than that of service contract for communal collection, but there was no significant correlation between service quality and productivity. The service quality was compromised since the capacities of the companies could not cope with the increasing waste generation even though higher productivity was achieved. The results suggest that companies' experience, scale of operation, and supervision capacity lead to high productivity. The productivity of private companies increases with increasing scale of operation and decreases beyond optimum scale of operation of 5 to 7 vehicles. The results of the factors such as route planning, operation scheduling, workers morale, and in-house maintenance influencing productivity of the companies were mixed and inconclusive.

Results on private sector involvement show that there has been increasing role of the private sector in solid waste collection service over the past two decades with more than 60% of the solid waste in the cities collected by the formal private sector. The practices of involving private sector have been a delineation of cities into zones, engagement of companies through competitive bidding or no bidding, and rendering of service with or without signed contractual agreements between companies and Local Governments. Notwithstanding the increasing involvement of the private sector, the involvement of households to pay for service provision, monitor service quality, and separate waste at source has been slow. Most of the people are aware of preventive measures to deal with illegal dumping and solid waste problems. The Local Governments and companies have not ensured customer-oriented service due to the slow pace of user charging and low cost recovery. A shift towards cost recovery through charging of all households on pay-as-you-throw (PAYT) basis for communal service and monthly fixed charges for kerbside collection (house-to-house) started first in

Tema and recently in Kumasi and Accra. The PAYT approach has potential for full cost recovery in low income communities, but in Accra monthly service fee was imposed on low income households. The fee-based collection in Accra is faced with problems of financing bins, unwillingness of households to register for collection, low payment rate by low income householders, illegal dumping by some households who refused to register, high disposal fee, and weak enforcement of registration and illegal dumping by the Assembly.

There were weak regulatory practices and non-adherence to contractual obligations, resulting in no incentives for full cost recovery and better service quality. The regulatory practices and 'capture' which affected service quality are non-competitive bidding and unsigned contracts. This normally results in prolonged period before upward review of collection fees and service charges, delay of subsidy payment, and no interest paid on monies delayed. The delay of subsidy payment (for more than a year) to the private companies does not provide incentive for private sector investment in new vehicles and does not enhance better service delivery. Results suggest that 'regulatory capture' exist in contracting since some companies had contracts signed and received regular payments whereas others did not. There were weak adherence to contract obligations, weak sanctioning of private companies for non-compliance, and weak enforcement of by-laws. The weak enforcement of legislation does not provide incentives for better service quality and full cost coverage. Regulation of prices by the Local Government, where service charges and collection fees levels are not sufficient to recover full cost of service, undermine the service quality of the companies. The major constraints of solid waste management were the inadequate capacity and financial resources of the Assemblies and regulatory capture by politicians, and the high risk of investment by the companies and the difficulty of getting funding for new standard collection vehicles. However, there is now a gradual well functioning system being put in place with the recent competitive bidding, signing of contracts, and city-wide user charging.

The conclusions drawn from the study are that: (1) there was no significant correlation between service quality and productivity, but there exist significant difference in service quality across cities, and that the service quality of house-to-house collection was higher than that of communal collection, (2) the analysis of the evolving involvement of the private sector show that there was zoning of the cities into solid waste collection zones and an increasing role of the private sector in solid waste collection service over the past two decades with each company operating in a defined collection zone, (3) despite the increasing involvement of the private sector, the involvement of households in paying for service provision, monitoring of service quality, and separating waste for recycle and recovery has been slow, (4) a shift from free service towards full cost recovery, through city-wide-based fee solid waste collection, where all households are charged on pay-as-you-throw (PAYT) basis for communal service and monthly flat charges for kerbside collection has potential for better service quality, (5) there were weak regulatory practices and non-adherence to contractual obligations, and these consequently provided little incentives for cost recovery and better service quality, (6) the findings suggest that companies' experience, scale of operation, and supervision capacity lead to high productivity, (7) the major constraints of solid waste management were the inadequate capacity and resources of the Local Governments (Assemblies), regulatory 'capture' by politicians, high risk of investment by the companies, use of over aged collection vehicles, low cost recovery and long delay in paying for the contracts, and inadequate monitoring of quality of service. This study concludes that the solutions to the problems of solid waste management hinge on adherence to formal rules, use of appropriate cost recovery mechanism for low income group, and restructuring of institutional arrangements to ensure bottom-up approach to user involvement, enforcement of legislation and capacity building.

De rol van de private sector bij het ophalen van vast afval in de stad
Prestaties, capaciteit en regulering in vijf steden in Ghana

Samenvatting

Dit proefschrift gaat over de rol van de private sector bij het ophalen van vast afval en over de vraag in hoeverre de capaciteit van de private sector en voorschriften van de lokale overheid de prestaties van de private sector beïnvloeden. De private sector is betrokken geraakt bij de publieke dienstverlening om problemen op het gebied van marktwerking en het functioneren van de overheid aan te pakken (van Dijk, 2008b). De publieke sector speelt een hoofdrol als afnemer (koper) namens de burgers door middel van subsidies en/of gebruikerskosten, terwijl de private sector steeds vaker optreedt als aanbieder (verkoper) van publieke diensten. Daarbij wordt de private sector gereguleerd door de publieke sector om problemen met de marktwerking die voortvloeien uit het feit dat de afvalinzameling een publieke dienst is te voorkomen. Betrokkenheid van de private sector is mogelijk een manier om marktwerking te behouden en om management- en technische expertise en private financiering te introduceren in de publieke dienstverlening. Hierdoor zouden de kosteneffectiviteit en de dienstverlening moeten verbeteren (betere kwaliteit van de diensten) (Cointreau, 1994; Bartone, 2001; van Dijk, 2008b). Betrokkenheid van de private sector (private sector involvement; PSI) biedt mogelijkheden om de efficiency te verhogen en de dienstverlening te verbeteren. Er is echter maar weinig onderzoek gedaan naar PSI bij het verwerken van vast afval in ontwikkelingslanden en dit onderzoek was niet gericht op het functioneren van de private sector. Het doel was niet om vast te stellen wat de voordelen van PSI zijn en welke factoren van invloed zijn op de efficiency en kwaliteit van de dienstverlening op bedrijfsniveau.

Dit onderzoek richt zich in de eerste plaats op de ontwikkeling van de betrokkenheid van de private sector bij het ophalen van vast afval in steden, en op de factoren die de verschillen tussen de prestaties van afzonderlijke bedrijven verklaren. De belangrijkste onderzoeksvraag is in welke mate de capaciteit voor en regulering van dienstverlening op het gebied van vast afval de prestaties van de private sector beïnvloeden. Om de omvang en oorzaken van problemen bij de dienstverlening vast te stellen en mogelijke beleidsinterventies om deze problemen aan te pakken voor te stellen behandelt het onderzoek de volgende vragen: Hoe heeft de betrokkenheid van de private sector zich ontwikkeld? Wat is het prestatieniveau van bedrijven (productiviteit en kwaliteit van de dienstverlening)? Welke interne en externe factoren verklaren het prestatieniveau?

Het onderzoek is uitgevoerd in de vijf grootste steden van Ghana (Accra, Kumasi, Tema, Takoradi, en Tamale). Deze steden zijn gekozen vanwege de snelle urbanisatie en uitbreiding. Er zijn kwalitatieve en kwantitatieve data verzameld bij vijf overheidsorganisaties en vijfentwintig particuliere bedrijven. De productiviteit van de bedrijven is beoordeeld op grond van secundaire gegevens over de bedrijven in de vijf steden. Er is bij een steekproef van 1500 willekeurig gekozen huishoudens een vragenlijst afgenomen in de deelnemende steden. De deelnemers aan het onderzoek konden per huishouden hun mening over de kwaliteit van de bestaande dienstverlening geven op een vijfpuntschaal. De gegevens van de huishoudens zijn geanalyseerd om de kwaliteit van de dienstverlening en de tevredenheid van de inwoners te bepalen. Het onderzoek besteedt ook aandacht aan de implicaties van de resultaten voor prestaties, reguleringsbeleid en duurzame dienstverlening.

Uit het onderzoek naar de prestaties van de private sector blijkt dat er een significant verschil was tussen de kwaliteit van de dienstverlening in de verschillende steden. De kwaliteit van de dienstverlening door franchisenemers (huis-aan-huisinzameling) was hoger dan die van bedrijven met een dienstverleningscontract voor collectieve inzamelsystemen, maar er was geen significante correlatie tussen kwaliteit van de dienstverlening en productiviteit. De kwaliteit van de dienstverlening verminderde omdat de capaciteit van de bedrijven niet toereikend was voor de toegenomen afvalproductie, ook al nam de productiviteit toe. De resultaten wijzen erop dat de ervaring van de bedrijven, de schaal waarop zij opereren en hun mogelijkheden voor supervisie tot hogere productiviteit leiden. De productiviteit van particuliere bedrijven neemt toe naarmate zij grootschaliger opereren en neemt af onder de optimale operationele omvang van vijf tot zeven voertuigen. Factoren als routeplanning, planning van de werkzaamheden, moreel van de werknemers en onderhoud door het bedrijf zelf hadden geen eenduidig en doorslaggevend effect op de productiviteit van de bedrijven.

Uit de resultaten blijkt dat de private sector de afgelopen twintig jaar een steeds grotere rol is gaan spelen bij het ophalen van vast afval. Ruim 60% van het vaste afval in de steden wordt opgehaald door de formele private sector. De betrokkenheid van de private sector is gepaard gegaan met de indeling van de stad in zones, en er was in sommige gevallen wel, en in andere gevallen geen sprake van openbare aanbesteding en van ondertekende dienstverleningsovereenkomsten tussen bedrijven en lokale overheden. Ondanks de steeds grotere rol van de private sector betalen huishoudens nog maar in beperkte mate voor de dienstverlening, en leveren ze nog slechts een geringe bijdrage aan het bewaken van de kwaliteit van de dienstverlening en aan afvalscheiding. De meeste mensen zijn op de hoogte van preventieve maatregelen om illegale dumping en problemen met vast afval tegen te gaan. De lokale overheden en bedrijven zorgen nog niet voor klantgerichte service doordat de kosten nog onvoldoende aan de gebruikers doorberekend worden en onvoldoende terugverdiend worden. Er is echter onlangs een steeds beter functionerend systeem ingevoerd met openbare aanbesteding, het tekenen van contracten en het in rekening brengen van kosten aan gebruikers in de hele stad. Een omslag naar kostendekking door alle huishoudens te laten betalen voor collectieve inzameling en afvoer van afval op basis van het principe 'de vervuiler betaalt' (pay-as-you-throw; PAYT) en door een vaste maandelijkse bijdrage te vragen voor het ophalen van afval aan de weg (huis aan huis) is begonnen in Tema en dit systeem is recentelijk ook in Kumasi en Accra ingevoerd. De PAYT-aanpak is potentieel volledig kostendekkend en is in Kumasi toegepast bij lagere-inkomensgroepen. In Accra moeten huishoudens met lage inkomens een maandelijkse bijdrage voor het ophalen van afval betalen. Het tegen betaling ophalen van afval in Accra functioneert niet goed doordat er problemen zijn met de bekostiging van afvalcontainers en huishoudens onvoldoende bereid zijn zich in te schrijven voor de inzameling. Ook betalen huishoudens met lage inkomens vaak niet en dumpen sommige huishoudens die weigeren zich in te schrijven hun afval illegaal. Verder is de afvalheffing hoog, en wordt de registratie van huishoudens niet goed gehandhaafd door de lokale overheid.

Er was aanvankelijk sprake van zwakke regulering en contractuele verplichtingen werden niet altijd nagekomen, waardoor er geen stimulans was om kostendekkend te werken en de dienstverlening te verbeteren, maar er wordt nu een steeds beter functionerend systeem ingevoerd. De kwaliteit van de dienstverlening verminderde doordat contracten niet altijd getekend werden, afvalheffingen en prijzen van de dienstverlening niet tijdig verhoogd werden, en subsidiegelden te laat uitgekeerd werden, waarbij geen rente werd vergoed over het te laat uitgekeerde bedrag. Het (ruim een jaar) te laat uitkeren van subsidies aan de particuliere bedrijven stimuleert de groei van de private sector niet en leidt ook niet tot betere dienstverlening. Uit de resultaten blijkt dat er sprake is van *regulatory capture* (het beschermen van de belangen van gereguleerde bedrijven door de regulerende instantie), omdat sommige bedrijven ondertekende contracten hadden en regelmatig betalingen ontvingen, terwijl dit bij andere bedrijven niet het geval was. Verplichtingen werden vaak niet nagekomen, zoals het opleggen van sancties aan particuliere bedrijven die zich niet aan de regels houden en het handhaven van plaatselijke verordeningen. Door de gebrekkige wetshandhaving wordt een betere dienstverlening en volledige dekking niet gestimuleerd. Prijsregulering door de lokale overheid waarbij tarieven voor dienstverlening en afvalheffingen niet volledig kostendekkend zijn en waarbij investeringen in nieuwe vuilniswagens niet terugverdiend worden, ondermijnt de kwaliteit van de dienstverlening van de bedrijven. Het afvalbeheer werd in belangrijke mate gehinderd door de ontoereikende capaciteit en financiële middelen van de gemeenteraden en regulatory capture van politici.

Dit onderzoek leidt tot de volgende conclusies. (1) Er was geen significante correlatie tussen kwaliteit van de dienstverlening en productiviteit, maar er is een significant verschil tussen de kwaliteit van de dienstverlening in de deelnemende steden. Daarnaast was de kwaliteit van de dienstverlening bij de huis-aan-huisinzameling hoger dan bij de collectieve inzamelsystemen. (2) Uit de analyse van de ontwikkeling van de betrokkenheid van de private sector blijkt dat de steden ingedeeld werden in zones voor de inzameling van vast afval. De private sector is de afgelopen twintig jaar een steeds grotere rol gaan spelen bij het ophalen van vast afval, waarbij elk bedrijf in een bepaalde inzamelingszone actief is. (3) Hoewel de private sector een steeds grotere rol speelt, blijft de bijdrage van huishoudens achter wat betreft het betalen voor de dienstverlening, bewaken van de kwaliteit van de dienstverlening en hergebruik van afval. (4) Een omslag naar kostendekking door in de hele stad een afvalheffing in te voeren, alle huishoudens te laten betalen voor collectieve afvalinzameling op basis van het principe 'de vervuiler betaalt' (PAYT) en een maandelijkse bijdrage in te voeren voor huis-aan-huisinzameling kan de kwaliteit van de dienstverlening verhogen. (5) Er was aanvankelijk sprake van zwakke regulering en het niet nakomen van contractuele verplichtingen, waardoor er geen stimulans was om kostendekkend te werken en de dienstverlening te verbeteren. (6) De resultaten wijzen erop dat de ervaring van de bedrijven, de schaal waarop zij opereren en hun mogelijkheden voor supervisie tot hogere productiviteit leiden. (7) De grootste belemmeringen voor goed afvalbeheer waren de ontoereikende capaciteit en middelen van de lokale overheden (gemeenteraden), te oude vuilniswagens, lage kostendekking, veel te late betaling voor de contractueel vastgelegde dienstverlening, gebrekkige bewaking van de kwaliteit van de dienstverlening, en regulatory capture van politici.

De conclusie van dit onderzoek is dat het probleem van het beheer van vast afval opgelost kan worden door zich aan de formele regels te houden, het juiste kostendekkingmechanisme te gebruiken voor de lagere-inkomensgroepen, en een institutionele herstructurering door te voeren om wetgeving te handhaven en capaciteit op te bouwen.

1 Evolving Management of Urban Solid Waste Services

1.1 Introduction

Many Developing Countries are still struggling with solid waste collection and management (Wilson, 2007), and Ghana is no exception. Solid Waste is waste generated and discarded as useless or unwanted from activities in homes, institutions, public and commercial places, and industries, though it is also a resource for reuse, recycle, and recovery (Tchobanoglous et al., 1993). Solid waste collection in developing countries, especially in Africa, is a real challenge to the public sector. Given the level of investment, the running cost of solid waste management, and the competing priorities (water, health, education, roads, and energy) of national governments, the public sector alone could not deliver the solid waste services. The private sector is partnering with the public sector to provide the needed resources for the solid waste service delivery. But for this arrangement to have impact on total service coverage and environmental cleanliness, it will depend on the formal rules and enabling environment (policies, legal and regulation) created and maintained by the local and central governments to provide the needed incentives for more investment, improved service quality, and regular price adjustment (OECD, 2000).

This chapter first provides the context of the study by examining solid waste problems and challenges in developing countries. Second, it looks at evolving public service markets and private sector involvement in public services with emphasis on urban solid waste collection in developing countries. Third, it describes and provides the context of the solid waste collection and disposal situation in Ghana, the research problem, and objectives. The rest of the chapter describes the relevance of the study and the structure of the dissertation.

1.2   Problems and Challenges of Solid Waste Collection in Developing Countries

Solid waste problems

Despite the private sector involvement in solid waste management in developing countries in the past two decades, there are still problems with solid waste management services. The problems in cities have become burdensome despite efforts being made by city authorities and governments (Onibokun and Kumuyi, 1999). The problems of solid waste such as inadequate service coverage, irregular waste collection, waste spill over from bins and storage containers, and lax attitude of people towards indiscriminate disposal on unauthorised places and waste littering are common in developing countries (Zurbrugg, 1999; Onibokun and Kumuyi, 1999; Oduro-Kwarteng et al., 2006). These problems eventually lead to public health impact, aesthetic nuisance, and environmental pollution. Because of the insufficient service coverage, the uncollected solid waste by the public sector or its agent is often dumped into drains, rivers and surrounding areas, or it is locally burnt or buried. These practices lead to considerable environmental pollution and degradation, and pose serious health risk to the population. These problems impair, in the long run, not only the quality of life of the urban poorer communities but also affect the welfare of the entire urban population. The impact of uncollected waste within cities in developing countries is enormous. Cities are faced with urban environmental health issues related to solid waste management (UN-HABITAT, 1996:146).

These waste problems are attributed to a number of causes. First, rapid urbanization poses a big problem to urban solid waste management. The rapid and unregulated urban growth and development in urban areas lead to a situation where solid waste services infrastructure lag behind the growth in population. The rapid rate of uncontrolled and unplanned urbanization in the developing countries poses great problems to sanitation and solid waste management for city authorities. The high level of urbanization in developing countries implies high quantities of solid waste generated. There are several negative impacts of rapid urbanization in developing countries which need to be mitigated.

Second, there is low cost recovery for solid waste service coupled with limited funds from central government general taxes in developing countries due to competing national priorities. Local authorities have difficulties in keeping pace with solid waste facilities development and in meeting the growing demand for solid waste services due to financial constraints. The low cost recovery is also attributed to lack of involvement of households in paying for the service. There are poor access roads and streets in some portions of urban settlements which make it difficult to collect waste in such neighbourhoods. The needed solid waste infrastructure and equipment are often inadequate due to limited funds or budgetary allocations from central government to the local government.

Third, political will and priority related to solid waste service and social prestige of waste workers are still very low, although most governments increasingly acknowledge solid waste problems as immediate and serious. The lack of priority, political will, and public sector commitment limit rapid and sustainable improvements in the solid waste sector in developing countries.

Fourth, the weak enforcement of by-laws for solid waste management also contribute to the lax attitude of the people towards indiscriminate disposal at unauthorised places, waste littering, and free riding. The enforcement of regulation by government officials appears to be weak and this may be due to lack of capacity, lack of resources and political will, and problems with the institutional set-up.

Finally, there are problems with physical and human capacity for solid waste collection and technology operations in some countries. The human capacity of the public and private sector organizations may not be adequate for solid waste management. The efficiency and effectiveness of solid waste collection require system analysis and optimization of operations. The inability to do this usually leads to inefficient use of time and resources, and which eventually leads to high cost of collection service, low productivity and poor service quality (Zurbrugg, 1999; Cointreau-Levine and Coad, 2000).

Challenges of solid waste services delivery

Developing Countries are faced with a number of challenges in solid waste management. First, the capacity of the local governments (LGs), to manage contracts and 'enable' the efforts of non-state actors to deliver efficient and effective service, is not up to the level required for the facilitating, regulating and monitoring roles of the local governments (Awortwi, 2003). The 'enablement' requires local governments to reduce direct service delivery but manage better and differently (Helmsing, 2002: 323). The enabling role involves facilitating efforts of others to perform responsibilities that have been transferred from state departments, setting policies and regulating private providers, and monitoring the performance of private providers and assessing their performance. The LGs face challenges of contract management, public opposition, and monopolistic behaviour and inefficiencies of agents (Awortwi, 2003).

The second challenge is how to enforce legislations enacted to improve and sustain solid waste collection and management service – increase cost recovery, stop indiscriminate solid waste dumping on unauthorised places, and avoid waste littering – through polluter-pay-principle or user charges and establishment of efficient and effective institutional structure to deal with the weak enforcement. Finally, the capacity challenge has to do with how the local governments could remove the inefficiencies from their direct service delivery. There is the need for capacity to optimise collection operations to take advantage of costs reduction and maximum output of waste collected.

1.3   Private Sector Involvement in Solid Waste Market

The delivery of public services has traditionally been carried out by the public sector. The increasing financial burden on the local governments and the inefficiency of the public sector (government failure) in developing countries necessitate the use of markets for public service delivery. However, markets where there is perfect competition with willing buyers and sellers do not work for public services that have externalities and information asymmetry. Solid waste collection service as a public good has externalities (negative environmental impacts) if people are excluded from the service. Solid waste collection cannot be provided through the market without regulation (legislation and incentives). The private sector is involved in solid waste collection due to market and government failures. There is also non-governmental organization failure, due to the over reliance on donor support to cover investment, operation, and maintenance costs. This means that the private sector failure (inefficiency) – under performance and inability to deliver the expected service quality – could occur if the needed policies, legislation, incentives, and government support are not given to it.

The extension of the market mechanisms of the New Public Management (NPM) to private sector involvement in solid waste collection services is still an emerging issue, especially in developing countries. Contracting out solid waste services to the private sector and charging for services rendered by the private sector are still faced with difficulties. Public services delivery such as water supply, sanitation and solid waste services have been failing in developing countries for a long time despite the NPM and decentralization of local service delivery to the local governments. The expected improvements in service delivery have often not been achieved (van Dijk, 2006). Obviously, decentralization alone was not enough to bring about improvements in service delivery, and therefore private sector involvement in public service delivery was introduced. The paradigm shift from public sector delivery of public services (solid waste service delivery) in developing countries to private sector provision began in the past two decades. Governments vigorously began to promote the private sector as a provider of services to improve service efficiency[1] and effectiveness[2] (Roth, 1987; Cointreau-Levine and Coad, 2000; Batley and Larbi, 2004), but the needed private finance and expertise to bring about the improvement are still issues, especially in developing countries.

In developing countries, different forms of Private Sector Involvement (PSI) have been suggested for achieving greater efficiency and effectiveness, to overcome the government failures in public direct service delivery – too many workers, not enough supervisors, few incentives for better performance and limited finance (Cointreau-Levine, 1994; Cointreau-Levine and Coad, 2000; Post et al., 2003). Private Sector Involvement (PSI) in solid waste collection in developed countries emerged in the 1970s, and since then there has been increasing private sector involvement in solid waste collection service in many parts of the world (Eggerth, 2005). By 1994, there were more than 10,000 private firms engaged in urban solid waste collection service in the United States, where more than 80 percent of solid waste was collected by the private firms (Cointreau, 1994). There is now PSI in all the elements of integrated solid waste management from collection, sanitary landfilling, recycling to resource recovery in the developed countries.

Private Sector Involvement in all sectors in developing countries has been slow especially in the Sub-Saharan Africa countries, although there is increasing private sector involvement (PSI) uptake in French-Speaking Africa (Li and Akintoye, 2003). By 1989, there was private sector involvement in solid waste collection in Latin American cities (Santiago, Buenos Aires, Sao Paulo and Caracas) with populations of 3.6 to 12 million (Bartone, 1991). The companies in these cities operated under service contract arrangements with the municipalities. The involvement of private sector in solid waste collection in most developing countries started gaining momentum in the 1990s. The World Bank advocated Private Sector Involvement in the 1994 World Development Report. Since then, the development

---

[1] *Service efficiency* is a measure of the use of available resources in the best possible way to produce output.
[2] *Service effectiveness* means the extent to which the objectives of the service are met.

partners have supported the drive for PSI in solid waste collection and management through capacity building and loans for provision of equipment. The number of private companies involved in solid waste collection keeps on increasing in developing countries, as in the case of Ghana and there is growing interest of the private sector in many developing countries. However, the presence of PSI in urban solid waste collection in developing countries has not been felt in terms of better service quality and total service coverage, and this may be due to a number of issues such as policy, capacity, regulation, legislation, and investment risk.

1.4 Does Private Sector Deliver Efficient and Quality Solid Waste Service?

The rationale for the Private Sector Involvement (PSI) in solid waste collection is to improve efficiency (reduce cost) and effectiveness of service delivery (service quality) through competition for the market – where private sector providers compete for a zonal monopoly to render service over a period of time – and to ensure that the environmental aspect of sustainable development is integrated into solid waste management. However, recent case studies of PSI in solid waste management in some developing countries – for example, in Kenya (Karanja, 2002; Mwangi, 2003), in Ghana (Obiri-Opareh, 2002; Awortwi, 2003), in Tanzania (Mbuligwe, 2004; Kassim, 2006) and in India (Post et al., 2003) – showed that there has been an increased coverage in some of the countries, but the service quality, efficiency and sustainability of private sector service delivery are still issues that require further studies to identify drivers for performance. The private sector still faces challenging issues of inefficiency and low service quality due to some factors of the enabling environment, inter-organisational arrangements, and how companies are run; and this is what this study seeks to identify.

Studies on performance of service providers often arrive at the conclusion that services delivery by private sector is associated with gains in effectiveness and service efficiency more than by municipal departments

(Cointreau-Levine, 1994; Cointreau-Levine and Coad, 2000; Post et al., 2003). Other authors argue that the results of private sector performance (efficiency gains) over public sector delivery showed that efficiency gains are mixed and that the debate on private sector efficiency gain over public sector is inconclusive (Donahue, 1989; Bel and Warner, 2008). The results from these studies showed that their explanatory factors are inconclusive, and therefore require further studies into other approaches and examples. The debate concerning the service quality and efficiency gains of the private sector delivery, and the explanatory factors will be reviewed in chapter 3. But it is worth mentioning here that the argument of the proponents is that the main reason for the efficiency gain is ownership and competition whereas the critics argue that there are more other factors which explain the efficiency gain.

Private sector inefficiency in developing countries may be due to a number of factors, and one of them is operational inefficiency due to weak capacity. Zurbrugg (1999) argues that the operational inefficiency of solid waste collection service delivered in developing countries are due to weaknesses in institutional arrangements (policies, legal, and regulations), deficient capacity of the public and private sector institutions involved, and the use of inappropriate technologies. It follows from this that operational efficiency of the service agent among other factors are necessary for private sector efficiency gains and improved performance.

There has been increased involvement of the private sector in solid waste management in many cities in developing countries (Post et al., 2003; Cointreau-Levine and Coad, 2000). However, despite the increasing interest in public-private-community partnerships, there is evidence that coverage and the needed improvements in environmental sanitation have not been achieved (Onibokun and Kumuyi, 1999; Oduro-Kwarteng et al., 2006). The solid waste collection coverage has not improved to the desired level in the developing countries, despite the paradigm shift from public delivery of solid waste services to private sector participation[3].

---

[3] *Private sector participation* is a partnership between the public and private sectors, where an arrangement that allows the private sector to participate in service delivery is formulated.

There is an argument that the private sector does not, in some cases, guarantee higher effectiveness and efficiency gains or reduce cost. Some studies suggest that the efficiency of private sector depends on the capacity of local government institutions to regulate and monitor performance of the private sector[4], and to recover cost (Obirih-Opareh and Post, 2002; Awortwi, 2003; Obirih-Opareh et al., 2004; Oduro-Kwarteng et al., 2006). Apart from regulation and performance monitoring by the public sector, which are external to private sector organisation, there are other internal factors which affect private sector performance. The effectiveness and efficiency of service delivery by private sector depends on a number of factors, which may be internal or external to private sector organisation.

The performance improvement of solid waste services in developing countries is daunting and one would wonder where things went wrong in the management and provision of the services. Although literature on technical, policy frameworks, implementation strategies, urban governance and institutional dimensions of waste management is large and growing, detailed analysis of internal and external factors to private sector organisation is needed to be able to apply measures that will increase efficiency and effectiveness in solid waste service provision in developing countries. The theoretical framework for this study is based on theories of markets and regulation of public services. Market as a process involves market actors (buyers and sellers), exchange mechanisms (transactions), object of exchange (services), industry demand and supply, and regulators. The market and how it is regulated determines the service coverage, efficiency, and quality. The theoretical framework will be developed further in chapter 3.

---

[4] *Private sector*- is part of the economy in which economic activity is carried out by private enterprise as distinct from the public sector.

1.5   Private Sector Involvement in Sustainable Solid Waste Management

The urban solid waste management in developing countries is faced with challenges of sustainability. A sustainable solid waste collection and management system encompasses a system that is environmentally, financially, and socially appropriate and acceptable, and meets the criteria of sustainable development – development that meets the needs of the present generation without compromising the ability of future generations to meet their needs (World Commission on Environment and Development, 1987). The three important interrelated aspects (environmental, financial, and social) of sustainability when met ensure that solid waste does not cause environmental pollution and public health hazards (Baud et al., 2004; Baud and Post, 2003).

Environmental sustainability requires that solid waste collection and disposal which imposes great burden on the environment and resources, be transformed into a closed-cycle system (closing the loop) restoring various natural cycles, thus preventing the loss of raw materials, energy, and nutrients. This means that waste reduction and minimization are integral part of solid waste collection through source separation and separate collection. Financial sustainability ensures that there is a more sustainable cost recovery approach than reliance on government subsidy financing. Full or partial cost recovery through user charges based on ability-to-pay reduces the financial burden on the government. Social sustainability of solid waste collection concerns providing services to all strata of society, regardless of income. Formal solid waste collection and necessary institutional arrangements ensure total service coverage where everybody is served.

The arrangements for Private Sector Involvement in solid waste collection, waste recycling and recovery in most developing countries are most of the time on 'trial and error' basis. Post et al. (2003) in their paper titled 'trial and error in privatisation: experiences in urban solid waste collection in Accra (Ghana) and Hyderabad (India)' point to the fact that the issue of sustainability cannot be taken for granted. They argue that the fate of private sector involvement depends largely on the local relationships and how the state shapes PSI. There is still only scanty empirical evidence concerning the strengths and weaknesses of different organisational arrangements in service provision or concerning the overall impact of private sector involvement or privatisation policies (Post et al., 2003). Although private sector involvement in solid waste collection has become extremely popular in developing countries, the sustainability in the long run is still not certain. Public service delivery is faced with market failure and government failure (see chapter 3 for details). Private sector is now involved in service delivery but is regulated, and therefore to ensure total service coverage in developing countries, sustainability is paramount for solid waste service delivery.

The factors that drive sustainability of solid waste management in developing countries are still not certain. There is no systematic evidence on the local conditions and the factors that promote sustainability. Studies that unfold and link local conditions with drivers for sustainability may be necessary for improving and sustaining private sector involvement in urban solid waste collection, to the extent that private sector would be interested in solid waste recycling, recovery, and sanitary landfilling in developing countries. There are few private firms involved in solid waste recycling, recovery, and sanitary landfilling in developing countries because private sector is still facing challenges with solid waste collection services. This study seeks to provide a link between the local conditions and the drivers for sustainable solid waste management in Ghana to provide lessons learnt for other developing countries.

1.6   Coverage of Solid Waste Collection in Ghana

Urbanisation and Solid Waste collection

Five cities (Accra, Kumasi, Tema, Takoradi and Tamale) were selected for this study. They are the five largest cities and also regional capitals with varied socio-economic activities. Ghana's urban centres with populations of more than 5000 are classified into three: small towns, medium-sized towns, and large towns/cities. At the top of Ghana's urban hierarchy are the large towns/cities with populations of 250,000 or more; examples of such cities are Accra, Kumasi, Tema, Sekondi-Takoradi and Tamale. According to the 2000 Population and Housing Census Reports, the urban centres of Accra, Tema, Kumasi and Sekondi-Takoradi account for about 44% of the total urban population, with Accra and Kumasi cities alone representing about 34% of the total urban population. The remaining 56% of the urban population is found in about 350 urban centres in the country.

Access and coverage of solid waste service provision
Figure 1.1 shows the proportion of households and the mode of solid waste collection and disposal in the five largest cities. According to the 2000 Population and Housing Census (Ghana Statistical Service 2002), the mode of waste collection and disposal in Accra showed that 21 percent of residents were served by agents – contracted waste collectors – and 63 percent (using public dump/communal collection) dumped waste into communal storage containers or unto ground at official designated dump sites where the waste is collected by ground evacuation using payload equipment and tipper trucks. Since waste emptied at some public dumps is not collected, less than 70% of the solid waste in Accra was removed from official public dumps and houses for final disposal at the designated disposal site. Waste dumped elsewhere includes some unofficial dumping sites scattered in the city, and others category include gutters, drains, streams, and lagoons.

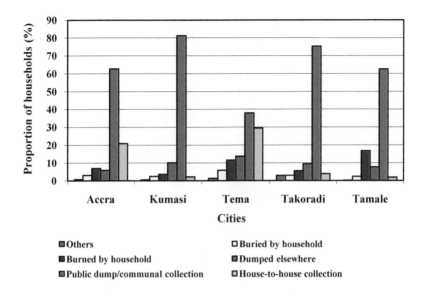

Figure 1.1 Mode of solid waste collection and disposal in 2000 in five largest cities
Source: Ghana Statistical Service (2002)

The proportions of people who used the public dump and communal collection – where waste is partly collected – ranged from 60 to 83% across
the cities. Since the public sector cannot cope with the ever-increasing volume of solid waste being generated in the cities, about 30% of the waste
generated in the cities are disposed of through recycling by the informal
sector, and by households through burning, burying, indiscriminate disposal (dumped elsewhere) especially in watercourses and drainage channels, which eventuality finds its way into the sea. The situation in all the
urban centres (with populations above 5000) shows that waste collected by
agents (both public and private through house-to-house method) as of 2000
was low (8.4%) and that using public dump (partly collected) and communal collection was high (67%) (Figure 1.2).

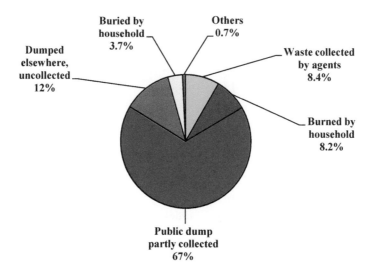

Figure 1.2 Solid waste collection and disposal in 2000 in urban towns and cities in Ghana. Source: Ghana Statistical Service (2002)

Collection rate of solid waste service provision

Collection rate is the amount of solid waste generated which is actually collected. Boakye (2005) showed that collection rates in 2004 across the cities ranged from 57 to 77, for more recent data see Chapter 5, Section 5.2.4. On the average about 68% of solid waste generated in the five selected cities was collected in 2004 (Figure 1.3). This was at the time PSI project financed by World Bank was being implemented in Ghana. Analysis of the collection rate and waste collected per population is useful for comparing PSI across cities. The analysis of waste collected by the companies, collection rates across the cities, and the evolving involvement of private sector will be explored in chapter 5. Chapter five will also cover the extent of private sector involvement in urban solid waste collection in Ghana so as to draw lessons learnt and conclusions for developing countries with similar characteristics as Ghana.

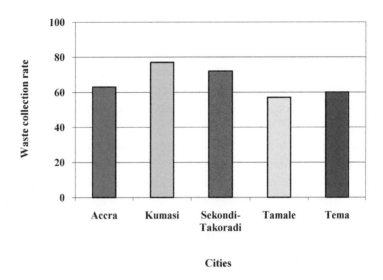

Figure 1.3 Solid waste collection rate (waste collected as a percentage of waste generated) in 2004 (Water Aid and WEDC, 2005; Boakye, 2005)

1.7   Research Problem

The solid waste sector in developing countries has not been able to provide adequate and sustainable solid waste service to the people. The main problem in cities in developing countries is inadequate and unsustainable access to solid waste services for the people and Ghana is no exception. The poor state of solid waste management – inadequate coverage, irregular waste collection, indiscriminate dumping, waste spill over from bins and storage containers, and waste littering – leaves much to be desired. The local governments which are responsible for solid waste management are faced with two problems. First, there is inadequate access to solid waste service especially in the middle and low income urban communities. Second, the solid waste services being rendered in the middle and low income urban communities are of poor quality and unsustainable. A significant proportion (about 30%) of the population in Ghana does not have access to solid waste services (Ghana Statistical Service, 2002; Oduro-Kwarteng et al., 2006), and those with access are concerned with better quality of the service such as frequent and reliable collection, avoidance of waste spill over from bins and storage containers, and quick response to customer complaints.

Public services such as water supply and solid waste collection have traditionally been organized by the public sector. With the introduction of PSI in urban solid waste management[5] in Ghana in the early 1990s due to the inability of the local governments to extend coverage and provide better services, it is expected that access to solid waste service and quality of the service would improve. Though some residents have benefited from private sector participation in terms of coverage and reliable service compared to when the services were delivered by the local governments alone (Obiri-Opareh, 2002), there are still issues of inefficiency of private sector delivery and service quality problems. Some residents still do not have access to solid waste services, and for some of those who have access to private sector collection service, the service is not reliable and effective – unreliable collection, overflow of bins and littering of waste (Awortwi, 2003). One would wonder if there are efficiency gains and effectiveness of service delivery by the private sector. Private companies show variations in productivity (Oduro-Kwarteng and van Dijk, 2008) and some have not delivered the expected outcomes – reliable service quality.

The problem addressed in this research is what regulatory (external) and firm level (internal) factors explain the variations in performance (productivity and service quality) of the private service providers involved in solid waste collection? Private sector involvement (PSI) in service delivery provides opportunities for improved service quality and coverage. However, PSI in most cases faces problems of policy implementation, weak

---

[5] *Management* refers to all activities that aim to minimize the impacts of municipal solid wastes on public health and the environment, including definition, characterization and measurement, storage, collection, transport, transfer, separation and resource recovery, processing and disposal.

regulatory frameworks, and inadequate private sector capacity due to cost recovery and subsidy issues. In order to establish the magnitude and causes of the inefficiency and service quality problems and possible interventions to these problems the following questions are addressed in this study: (1) Has private sector and household involvement evolved to ensure sustainable service delivery? (2) What is the level of company performance (productivity and service quality? (3) What factors explain the level of performance? This research seeks to identify the drivers for performance by examining the performance of private companies in five cities in Ghana.

1.8   Research Objectives

Private sector involvement in urban solid waste service delivery in Ghana is still at the infant stage with teething problems which require detailed analysis of performance of the service providers and the factors which explain their performance. Using Ghana, a developing country, as a case study the research issue to be addressed is how to explain private sector performance in solid waste collection and identify the factors which explain the differences in performance among companies involved in urban solid waste collection and management.

> *The goal of this research is to examine the evolving involvement of private sector in urban solid waste collection, and the factors explaining differences in performance (productivity and service quality) of solid waste companies.*

The specific objectives of the study are:

1. To explore the evolving practices of involving private sector and households in solid waste collection for sustainability of service delivery.
2. To analyse the productivity and service quality of private companies involved in urban solid waste collection service.

3. To examine and identify the internal and external factors which explain the differences in productivity and service quality of private companies.

The hypotheses for this research are discussed in chapter 3 and concern the relationships between performance (productivity and service quality of private companies) on one hand and the factors that explain the differences in performance on the other. To investigate the performance differences and the theoretical relationships, comparative analysis between companies in the same city and across the five different cities were carried out using both quantitative and qualitative methods.

1.9   Relevance of the Research

Studies on solid waste management in developing countries did not explore the influence of private sector capacity and local government regulation on private sector performance. The study involving private sector performance, capacity and regulation in solid waste management in developing countries is least explored. Studies on local governance of solid waste services (Awortwi, 2003) and impact of decentralization on solid waste management (Obiri-Opareh et al., 2004; Obiri-Opareh, 2002) conducted in Ghana focused on the policies for urban solid waste management. These two studies in Ghana did not focus on the link between private sector performance and regulation by contract.

   Studies conducted on solid waste management in other developing countries, for instance Kenya (Karanja, 2005; Mwangi, 2003; Rotioch et al., 2006), and Tanzania (Mbuligwe, 2004; Kassim, 2006) focused on service quality, institutional/governance issues, and the public and private relationship, but did not explore empirical evidence of the link between performance and its determinants and what could be done to address solid waste problems and improve urban solid waste management. These studies did not test empirical relationships between private sector capacity and productivity, but reveal solid waste problems and policy issues in urban solid waste management. The empirical relationships of theoretical issues influencing performance of service providers with testable hypotheses were not the focus of such studies (van Dijk and Oduro-Kwarteng, 2007), and will be taken up in this study.

There is little understanding of linkages between performance of the private sector and the regulation and capacity factors necessary for improved service delivery. To better understand the linkages between private sector performance and the different factors influencing performance, a look at the solid waste market, using four theoretical issues relevant for practice and policy, is necessary. First, what contractual relationship and mix of service (type of services) exist and are being operated by the private actors in the solid waste market? Second, does capacity ensures high performance for solid waste delivery? Third, what kind of household involvement and cost recovery mechanisms exist? Fourth, what kind of regulatory framework exists for regulating access to information, tariff/user charges, and quality of service in the solid waste market? The empirical evidence of the linkages between these four theoretical issues and the performances of the private actors will thus, provide an understanding of the internal and external factors that drive performance or act as barriers to performance.

The relevance of assessing performance of service providers involved in solid waste management in developing countries is an emerging issue. There are no standards or well defined performance targets for assessing performance of urban solid waste providers, and therefore different approaches are used by researchers. The performance data from this study could provide the basis for future performance monitoring of the private sector companies.

The output from this study will be useful for improving solid waste management service delivery to ensure effective and efficient solid waste service delivery. Specifically, the output will be useful for environmental sanitation policy formulation, performance monitoring, and future benchmarking to enhance efficient and effective service delivery. What is new in this study for Ghana is the effort to understand the variations in performance (productivity and quality of service) in the public-private institutional

arrangements. The study will provide an understanding of the linkages between the performance of the private service providers and the regulatory
arrangements necessary for improved service.

1.10   Organisation of the Thesis

This dissertation is organised into nine chapters. Chapter 1 introduces the
thesis and explains the evolving management of urban solid waste collection. This chapter presents the context of the study and research objectives.
Chapter 2 reviews the definitions and concepts of solid waste management
and the Ghana Solid Waste Sector. Chapter 3 reviews the theoretical and
conceptual issues of Private Sector Involvement in Solid Waste Collection.
The theoretical framework for analysing the capacity and regulation of the
private sector is also presented. Chapter 4 presents the research methodology for the study. Chapter 5 presents results of survey on the evolving involvement of private sector in solid waste collection in Ghana as well as
households' awareness and involvement in solid waste collection. Chapter
6 presents results on productivity and service quality in solid waste collection. Chapter 7 presents results on the private sector capacity for solid
waste collection service provision. It focuses on the scale of operations of
private sector companies and capability of companies. Chapter 8 presents
results on regulation of private sector in solid waste collection. It covers
contracts and financing arrangements (unit costs, user charging, pay-as-
you-throw charging, and cost recovery of collection service) and regulatory arrangements and controls by contract. Chapter 9 is the conclusion
and implications of findings. It presents the theoretical and practical conclusions from the study.

2 Institutional and Regulatory Context of Solid Waste Management in Ghana

2.1 Introduction

This chapter reviews the contextual issues of institutional (actors and roles) and regulatory arrangements for solid waste collection in Ghana. The fate of thriving private sector involvement (PSI) in service provision is shaped by the state and local relationships (Post et al., 2003). The role of the private sector and its impact on total service coverage and environmental cleanliness, depends on how formal rules and enabling environment (policies, legal and regulation) are created and maintained by the local and central governments, to provide the needed incentives for more investment, improved service quality, and regular price adjustment (OECD, 2000). Every organisation is affected by its external environment and activities of other actors. The customers (households) and institutions external to private sector and the frameworks for service delivery bring to the fore external factors in the external environment to the private sector. Even though there is little the private sector can do to change the external environment, it can adapt better to it and develop appropriate strategies to deal with the issues (Lusthaus et al., 1999).

This chapter is divided into five sections. Section 2 reviews the elements of solid waste management and the practices in Ghana. Section 3 reviews the institutional structures and arrangements for solid waste management. In Section 4, the policies, legal, and regulatory frameworks which are the external environment within which the private sector operates are reviewed. These frameworks provide the contextual issues and background for the study. The last section draws conclusion from the review on the contextual issues.

2.2   Elements and Actors of Solid Waste Management

Solid Waste Management and Role of Actors in the Service Chain

Solid Waste Management (SWM) is defined as managing the processes involving solid waste collection, treatment and disposal of waste generated in households, commercial and business establishments, institutions, and non-hazardous industrial process waste (Tchobanoglous et al., 1993). Tchobanoglous et al. (1993) grouped the direct activities of solid waste management into six functional elements: (1) waste generation and characterization, (2) on-site storage and handling, (3) collection, (4) transfer and transport, (5) separation processing, treatment and resource recovery, and (6) final disposal. These functional elements require planning and management in order to achieve high quality of service. Schubeler et al. (1996) reported that agencies responsible for SWM often pay too little attention to integrated management approaches based on adequate information systems, management approaches, methods, and techniques. The procedures and methods developed for management (planning, operations, monitoring) may be different in the various organizations, and managers responsible for solid waste management take control of all the necessary management functions and are responsible and accountable for discharging the roles of the organisation in a cost effective manner.

The activities of service provision may be concentrated in one organization or fragmented over multiple organizations. The fragmented processes in the value-chain are shown in Figure 2.1. The solid waste service provision is fragmented into four stages and different organizations or actors are involved in the provision of services. These fragmented stages of activities in the value-chain require different actors and different arrangement for service delivery. These stages are: (1) waste storage and drop-off stage and it is the responsibility of the householders, and service provider may provide primary collection, (2) collection, transfer and transport stage is the responsibility of the service provider who may be a private sector organisation or a public organisation, (3) the actors involved in the reuse, recycle, recovery (processing and treatment) stage are the householders, itinerants buyers, waste pickers and recycle companies and, (4) final waste disposal in sanitary landfill or control dump site is managed by private or public organisations. The key elements to be distinguished in chain approach (from source separation of solid waste to final treatment and possible reuse options) are the different stages in the solid waste process, roles of different actors both formal and informal, relationships of actors in service delivery, economics of each stage, and the improvement in service delivery through policy and regulatory frameworks (van Dijk and Oduro-Kwarteng, 2007).

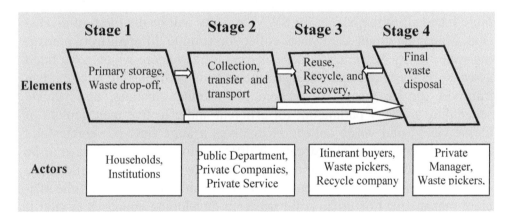

Figure 2.1 Stages in solid waste management (author's construct)

Developing countries are far behind the developed countries in the provision of solid waste services, since in most developing countries the introduction of environmentally acceptable sanitary landfilling and the integrated approach towards SWM are now being considered. Developing countries and Ghana in particular are lagging behind, perhaps, due to the prevalence of market failures and the inability of governments to provide these services. Solid Waste Management (SWM) aims to minimize the impact of municipal solid wastes on public health within communities and the environment, to promote environmental hygiene and cleanliness by preventing environmental pollution, and to recover and recycle valuable resources to minimize waste and disposal cost as well as stimulate local employment opportunities. This study attempts to explore the households' awareness of reuse, recycling and source separation of solid waste.

Urban solid waste collection services

For the purpose of this study, the solid waste collection service which takes more than half of the budget for SWM in a city will be the focus of discussion. Managing urban solid waste collection is quite different from managing water utility, sewerage system or road infrastructure which have fixed asset. Solid waste collection is a service and can be likened to public urban transport service where trip generation, routing planning, and vehicle scheduling are very paramount for efficient and effective management of the service. Solid waste collection involves a large fleet of vehicles and solid waste generation and storage at different locations which have to be managed. The performance of the companies involved in urban solid waste collection has effects on the environmental health. It follows that the efficient management of urban solid waste in developing countries is vital if environmental cleanliness and public health are not to be compromised.

Solid waste is defined as waste arising from human activities and normally solid and is discarded as useless or unwanted, though it is also a resources for reuse, recycling and recovery (Tchobanoglous et al., 1993). In other words, solid waste includes all solids or semi solid material that is of no value to retain in a given setting and is discarded. Urban solid waste encompasses domestic waste, street waste (sweepings from streets, parks, and playground), commercial waste, institutional waste as well as industrial waste. The U.S. Environmental Protection Agency defines urban solid waste to include food waste, yard waste, durable goods, containers and packaging, and miscellaneous inorganic waste from residential, commercial, institutional, and industrial sources. Urban solid waste excludes all categories of hazardous waste from industries, agricultural activities, sewage sludge, and medical waste.

Solid waste collection has evolved over the years from collection of unsegregated waste and disposal on dumping grounds to collection of source separated waste streams through formal and informal service providers. The management of the processes involved which was traditionally public has now become public-private-community provision and partnerships arrangements. The management approaches depend on the culture and society where solid waste is generated and their perception about solid waste.

There are four modes of solid waste collection service[6] depending on the income levels of the people, housing types and the level of service required. First, the communal[7] collection mode is rendered in low income areas. The householders discharge their waste into communal storage containers at transfer stations or designated locations and collection vehicles pick up the containers full of waste at frequent intervals. Second, block collection mode is used where there are large apartment residential buildings. For this service, collection vehicle travels along a predetermined route at specific intervals (every 2 to 3 days) and stops at selected locations. The householders bring their waste bins upon hearing a bell sound and hand them to the crew who empties the bins and gives them back to the householders.

Third, the kerbside collection mode is rendered in middle and high income areas. The collection crew collects bins and bags of waste which are

---

[6] *Collection service* involves the collection of solid waste from households in the municipality or from communal storage point.

[7] *Communal collection*- A system for solid waste collection in which individuals bring their municipal solid waste directly to a central point, from where it is collected.

deposited at the kerbside on fixed days (e.g. 2 specific days in a week) when collection takes place. The householders leave their bins at the kerbside and collect them later in the day. Finally, the door-to-door or house-to-house[8] collection mode is rendered in middle and high income areas. The collection crew enters each premise, takes out the bin and sends it back after emptying the waste into collection vehicles. No bins are left outside household premises. The residents served by the kerbside and house-to-house collection use standard bins to store waste. All the modes of collection are practice in developing countries but the common practice is the communal collection. The kerbside, house-to-house and the communal collection services are practised in Ghana. Solid waste services concern the link between customers and the service provider, and it is an integral part of solid waste management. The next section discusses the institutional arrangements for solid waste service provision.

2.3   Institutional Structures for Solid Waste Management in Ghana

2.3.1   Institutional Structures for Local Service Delivery

This section explains the institutional set-up for solid waste management in Ghana. It provides the context for a more systematic analysis in chapters 5 and 8, and better understanding of the relationships between the dimensions of regulatory arrangements and the quality of service delivery. The institutional arrangements for solid waste collection cover a range of dimensions such as mode of delivery, role of actors, and financing arrangements. These arrangements characterize and provide framework for the role of the private sector, public sector and the households in the various

---

[8] *House-to-house collection*- Method of collecting domestic solid waste in which the householder is responsible for putting the waste (in a container) outside his property at the curb- or roadside at specified times for collection. The container should generally be kept within the property.

stages of the waste management. There has been a shift, over the years, in solid waste service provision from central government to sole provision by local governments and now the involvement of private sector.

Central Government Provision of Solid Waste Collection

Table 2.1 shows the historical developments in solid waste service delivery from 1957 to date. The provision of public services in the colonial and post-colonial periods was done by the central government. After independence in 1957, the Public Works Department (PWD) was established to provide sanitation services in urban and rural areas. In the late 1970s, the public service provision deteriorated badly in the cities, and the central government (CG) could hardly manage the wide range of services (Awortwi, 2003). With worsened economic conditions in the early 1980s, there were insufficient funds for acquisition of capital equipment, resulting in regular breakdown of vehicles, plant and equipment. This situation reached a crisis level in 1985 (Benneh et al., 1993:38; AMA/Colan Consult, 1995). Most of the people had no access to solid waste collection services at all and dumped their waste anywhere they wanted.

Solid Waste Collection by Waste Management Departments Alone

In 1985, a Waste Management Department (WMD) was established in Kumasi with support from UNDP. Similar arrangements were made in the other cities (Accra, Tema, Takoradi, and Tamale), equipped and funded by the German Agency for Technical Co-operation (Deutsche Gesellschaft fur Technische Zusammenarbeit GmbH – GTZ) prior to the establishment of the Local Governments. The waste management problems were tackled more seriously through regular replacement of worn-out vehicles and equipment, and provision of funds for capital and recurrent costs, but beneficiaries did not pay service charges to WMDs. After the GTZ project, the city authorities could not sustain and maintain the equipment to enable the WMDs to operate satisfactorily. In Accra for instance, the collection of service charges from few wealthy and medium-income neighborhoods did not cover even half of the operational costs (AMA/ Colan Consult, 1995:35). In Accra, service coverage, which as a result of GTZ assistance had increased to about 80%, began to fall again, and by early 1990 it had fallen to 45% because of limited financial resources for maintaining the trucks (AMA/Colan Consult, 1995). Furthermore, there was no effective monitoring of activities and cost. Solid waste collection and sanitation management problems became headaches for the city authorities, and the residents perceived the WMDs as inefficient and ineffective (Agbe-mabiawo, 1996; Karikari, 1999).

Table 2.1 Historical Development in Solid Waste Collection

Year	Management mode
1957	Public service provision by Public Works Department (PWD)
1985	Waste Management Departments (WMD) were established in all the major cities, equipped and funded by the German Agency for Technical Co-operation (GTZ).
1988	Provision and delivery of basic public services were decentralized to local government (Adarkwa, 2005).
1990	In the early 1990s, there was policy shift towards private sector-led development, led to contracting out and franchising the solid waste collection services to the private sector in Accra and Tema
1995	Private sector involvement (PSI) in solid waste services in four of the selected cites (Tema, Takoradi, Kumasi, Tamale) was supported as component of Urban Environmental Sanitation Project (UESP-World Bank funded), (private sector was to collect 40% of waste),
1999	Policy shift in public private partnerships arrangements (private sector to collect 80% of waste),

In 1988, the provision and management of some of the basic public services, including solid waste and sanitation were decentralized to local governments (Adarkwa, 2005). Following decentralization of basic services, the Local Governments (LGs) set up service units and offered the services virtually free of charge, in the same way that the central parastatals had done in the past. The ability of LGs to deliver solid waste and sanitation services deteriorated sharply due to weaknesses in service operations, management, planning, attitudes of residents towards waste management, lack of enforcement of solid waste and sanitation by-laws, and unmotivated waste management personnel (Awortwi, 2003).

Private Sector Involvement in Solid Waste Collection

The private sector-led developments in the developed countries and the structural adjustment programme in the 1980s in Ghana made critics to question the service delivery role of Local Governments. In 1990 the LGs of Accra and Tema took the initiative and subsequently started contracting-out franchising of solid waste and sanitation services to the private sector. However, there was no provision in the legislation that allowed other actors than the LG to provide these services until 1992.

Private sector involvement in urban solid waste management was an integral part of Urban Environmental Sanitation Project (UESP) which was funded with World Bank loan and implemented in Tema, Takoradi, Kumasi, and Tamale from 1995. In the plans to involve the private sector, it was envisaged that in the short to medium term, LGs would provide about 60% of the basic services, with 40% being in the hands of the private sector. LGs were also required to set tariffs at realistic and economically viable levels, with due allowance for recurrent cost recovery and depreciation of capital investments, and to build private sector capacity through access to investment funds. The move from direct provision to contracting-out implied that LGs had to completely re-organize, reorient their management and regulatory apparatus and strengthen their human capabilities (Awortwi, 2003), in order to facilitate, regulate, and monitor the private sector provision of solid waste management services.

Institutional Arrangements for Solid Waste Collection

Figure 2.2 shows the relationships and roles of the stakeholders (and institutions) in the institutional arrangements for solid waste collection. The key institutions concerned with the implementation of the solid waste collection aspect of the Environmental Sanitation Policy are the Ministry of Local Government and Rural Development (MLGRD) and the Local Governments (Assemblies). MLGRD formulates solid waste and sanitation policies and also provides oversight role to the assemblies and gives subsidies for provision of SWC services.

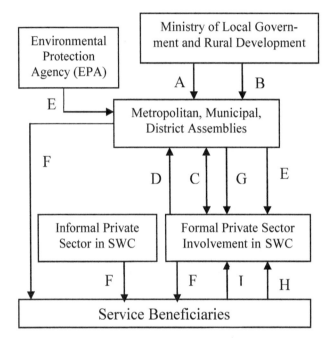

A - Solid waste policies formulated for the Assembly to implement
B - Provision of funds for solid waste collection activities
C - Submission of reports on activities of private companies and access to data from Assembly
D - Payment of charges to the Assembly by private companies.
E - Supervision and monitoring of activities of service providers
F - Rendering of services to service beneficiaries by the Assembly or private companies
G - Release of funds to private companies for services rendered
H - Payment of user charges to private companies by service beneficiaries
I - Complaints about service quality by beneficiaries,

Figure 2.2 Relationships and roles of stakeholders in the Institutional Arrangements for solid waste collection
(Source: Oduro-Kwarteng (2009) 34th WEDC Conference)

The Ministry monitors the activities of local governments (Metropolitan and Municipal Assemblies) and passes directives as determined by state legislation. Waste Management Departments of the Metropolitan, Municipal and District Assemblies (MMDAs) have the responsibility for solid waste management. The waste management functions were transferred from the Central Government to the Local Governments, as part of the decentralisation process in Ghana in 1988. The Metropolitan Assemblies receive much of their revenue directly from the Central Government.

The Metropolitan, Municipal and District Assemblies have a number of functional departments and one of the departments is the Waste Management Department (WMD), which is responsible for the liquid and solid waste collection and disposal, public cleansing, education of the public on waste management and supervision of activities of private contractors engaged by the Assemblies. The sub-metro cleansing or sanitation officers are responsible for solid waste collection monitoring. The health inspectors of Environmental Health Department (EHD) are expected to regulate and monitor the quality of service delivery and sanitary conditions, and sanction possible offenders.

2.3.2 Decentralisation of Public Services and Governance

The meaning of decentralization not only differs from one country to the other but even within sectors (Schwartz, 2006a). Decentralization signifies different meanings in various context and therefore terms such as "political decentralization", "administrative decentralization" and "financial decentralization" are used in literature (Adarkwa, 2005). Decentralization is defined as the transfer of authority (power by law), transfer of responsibility (roles and tasks) for specified public functions, and transfer of resources from higher level of (central) government to a lower level or from higher level of government to a subordinate or quasi-independent government organisations or private sector (Rondinelli, 1999, as cited in Laryea-Adjei, 2006). Four forms of decentralization are distinguished in literature: political decentralization, administrative decentralization, spatial decentralization, and market decentralization.

Three types of decentralization derived from combinations of these forms of decentralization are distinguished. First, *deconcentration* is one of the types of decentralization which is the redistribution of responsibilities only within the central government but not transferred to the local government. It is considered as the weakest type of decentralisation. Second, another type is *delegation* which involves the transfer of responsibility of decision-making and administration of public functions (for planning, implementing, investing) to (semi-) autonomous organisations that are not under the direct control of the government (or not wholly controlled by government) but ultimately accountable to it. This is more extensive form of decentralisation than administrative deconcentration. Last but not least is *devolution* which is the transfer of authority for decision making, finance and management from central government to quasi-autonomous local government units.

Public Service Delivery and Decentralisation in Ghana

In Ghana, two variant forms of decentralisation are currently being implemented (Adarkwa, 2005). First, there is the devolution of major political and administrative responsibilities from central government to Assemblies for local governance and general development. Devolution in Ghana has to do with restructuring technical set-up, political and administrative institutions and mandates to deliver services particularly at the local level. The delivery of solid waste management services has been decentralized and is now the mandate of the local government. The decision making on public or private delivery and the enforcement of regulation are the responsibility of the local governments (Assemblies).

The other which operates alongside the first, is the process of administrative and technical deconcentration of the planning and delivery or provision of specific services such as water and sanitation, health, education, and agriculture from Ministries, Departments and Agencies (MDAs). Most of the government agencies fall under this category, and an example is community water and sanitation agency and all the commissions. The deconcentration involved the re-arrangement of the location of key personnel and where their functions are performed.

Obiri-Yeboah et al. (2004) argue that the decentralisation effort in Ghana – to strengthening the MMDAs and creating Sub-Metropolitan Assemblies within the Metropolitan Assemblies – has performed far below expectation due to impediments associated with the decentralisation. Obiri-Yeboah et al. (2004) distinguished two factors which explain the flaws of the reform. The first refers to the failure of the decentralisation policy (on devolution) to empower the people through the Assemblies, to restructure the government machinery and reform local governments to take responsibility for local administration and development and to make public administration more efficient and responsive to local needs. The bottom-up approach to user involvement in local service delivery is an issue affecting local development, especially solid waste management and sanitation services which are core responsibility of the local governments.

The second refers to the lack of restructuring of the allocation of resources and reassignment of functions and responsibilities to the three levels of government (national, regional and district). There is insufficient transfer of resources from the central government to the Assemblies in order to match the transfer of responsibilities. For example the WMDs do not have financial autonomy and continue to depend on the highly politicised local authorities. The weakness of resource allocation is because of the administrative and technical deconcentration of the planning and provision of specific services. A lack of financial independence (solid waste collection is funded partly by central government resources) and the fact that there is still no composite budget for the Assemblies compound these problems.

Though decentralisation has brought major institutional changes such as making the district assemblies the pivot and focal point of planning and development, the administration of the Assemblies in Ghana continues to be weak since the central government ministries, departments and agencies (MDAs) continue to exercise tremendous authority and make spending decisions in a highly centralised and deconcentrated fashion. The local governments have little authority to make strategic decisions on infrastructure services. They have methods of raising the tax revenues but revenue generation has always been problematic due to corruption, politics, and lack of capacity. The reduction in central transfers and the deepening financial crises at local level compelled central governments to surrender to pressures to set up central grant systems (Assemblies' Common Fund).

Obiri-Yeboah et al. (2004) point to gaps between formal rules and actual practices in private sector involvement due to inadequate fiscal transfers from the central government to decentralised local governments to match increased responsibilities. There were problems such as: weak internal revenue mobilisation capabilities of the local governments, the lack of effective urban governance and its effects on service provision, and inadequate attention to environmental concerns in the development programmes. There is the tendency of the central government to interfere in purely local matters despite the decentralisation policy.

The LGs with the requisite capacity will be able to manage agents to benefit from the presumed advantages of contracting out service delivery. The LG capacity requirements include: developing legal framework in which other actors operate; creating a competitive contract award process; developing realistic financial incentives for agents; setting effective monitoring and compliance system; and improving organizational and management skills (Awortwi, 2003).

The use of markets and contracts within the context of decentralization of service delivery is an emerging process in developing countries and is not yet fully understood. The power relations in the decentralization processes and the new role of the local governments to use market mechanisms (contracting-out) are least explored in developing countries. The review has shown that the framework of decentralization has changed the actual role of the national and local governments and a more important role for the private sector emerged.

2.4 Legal, Regulatory and Policy Frameworks of Solid Waste Management

The local context of legal, regulatory and policy frameworks is believed to have significant influence on solid waste collection in developing countries to some extent. This section explains the situation in Ghana to provide the context for exploring the regulation and evolving involvement of the private sector in solid waste collection.

2.4.1 Legal and Regulatory Frameworks

The Central Government confers local authority status (Metropolitan, Municipal, and District Assemblies) on any urban area in line with the local government law which was amended in 1993 (Act 462) to replace the previous act enacted in 1988. Under current legislation in Ghana, the Central Government has a number of controls over Metropolitan, Municipal, and District Assemblies (MMDAs). The Central Government appoints chief executives to run MMDAs, and the Minister for Local Government approves the Assemblies' by-laws, budgets, and proposals to tap new sources of revenue or increase existing taxes. The Central Government also occasionally issues directives affecting the Assemblies. Of crucial importance is the fact that much of the revenue of the Assemblies comes directly from the Central Government. However, the MMDAs have full constitutional responsibility for management of sanitation and solid-waste collection and disposal. The central-local government relationship is believed to have implications for the governance of solid waste collection in the cities.

The criminal code on public nuisance was the only national legislative instrument on environmental sanitation prior to 1992 and it particularly sought to address indiscriminate disposal of waste at public places and prosecution of offenders. The Criminal Code of Ghana, by-laws of the local authority (Assembly) and national legal regulatory instruments have been promulgated and enacted into law and enforceable in a law court. The Criminal Code of Ghana, 1960 (Act 29) provides that whoever places or permits to be placed, any refuse, or rubbish, or any offensive or otherwise unwholesome matter, on any street, yard, enclosure, or open space, except at such places as may be set apart by the local authority or health officer for that purpose commits a punishable offence. The law seeks to ensure that residents take responsibility for the streets in front of them as well as their premises.

There are model by-laws, regulatory and legislative instruments for solid waste management. The laws in Ghana make it mandatory that the owner of the waste generated in the city is the Assembly, and therefore has the right to collect, recycle and dispose of the solid waste. The National Building Regulations, 1996 (LI 1630) stipulates that a building for residential, commercial, industrial, civic or cultural use shall have a facility for refuse disposal. It went further to state a requirement that each dwelling unit shall have a standardised dustbin or other receptacle approved by the Assembly in which all refuse generated shall be stored temporarily. It provides for transfer stations to be located within reach and preferably protected from rain and the prevention of spreading, pest infestation, and scavenging activities.

The other relevant legal provisions for the Assemblies to function are: Procurement Act and Assembly Tender Board regulation, By-law of the Assemblies, Environmental Protection Agency Act, 1994 (Act 490); Environmental Assessment Regulations, 1999 (LI 1652); National Building Regulations, 1996 (LI 1630); Town and Country Planning Ordinances, 1944 (Cap 84); Vaccination Ordinance Cap 76; Food and Drugs Law 305b (1992); and Mortuaries and Funeral Facilities Act, 1998 (Act 563). All these legal frameworks have provisions for sanitation and solid waste management and therefore require the effort of the public sector institutions to implement and enforce the laws for effective solid waste management.

The Local Government Act 462 gives the Ministry of Local Government (MLGRD) three responsibilities for SWM: (1) Policy and planning, (2) legislation on SWM, and (3) regulation, monitoring and enforcement of SWM activities. The implementation of these functions is the responsibility of the local government (Assemblies). City authority as legal entity has powers conferred on it by the Local Government Act 1993 (Act 462) to promulgate by- laws to govern and regulate solid waste management, sanitation, cleansing and abatement of nuisance in the city. Waste Management Department (WMD) is responsible for waste collection and transport operations, management of disposal (landfill) sites, repair and maintenance of waste vehicles and equipment. The solid waste problems come about as a result of the gaps in effort of implementation of polices and legal provisions. The next section reviews policies for solid waste management in Ghana.

2.4.2 Policy Framework

Policies on Solid Waste Service Provision

Policy is defined as the plan of action, set of procedures, rules and allocation mechanisms that provide the basis for programmes and services (WEDC, 2003). This definition covers both written policy documentation and the policy context, legislation, regulations, attitudes and assumptions that support that documentation. In practice, the term policy is commonly used to mean the document which outlines the intentions, decisions, and the course of action adopted by government towards achieving a goal. Policy as an overall statement of intent gives indication of how that intent can be turned into reality (WEDC, 2003). This research will focus on the extent to which policies are implemented.

The environmental sanitation sector is guided by the National Environmental Sanitation Policy published first in May 1999 and revised in 2008. The policy was revised to update its contents to meet current development objectives and address aspirations of sector actors after eight years of slow implementation with very little impact (MLGRD, 2008). The environmental sanitation policy, as a major determinant for improving health and standard of living in Ghana, is focused on seven key areas in Box 2.1. The policy reflects the changing context of national and international development priorities (Ghana Poverty Reduction Strategy, Millennium Development Goals, and New Partnership for African Development – NEPAD).

Box 2.1 Focus Areas of the Environmental Sanitation Policy

The new policy is specifically focusing on the following key areas:

Capacity development. This covers all issues of institutional development including human resource and institutional strengthening.
Information, Education and Communication. This entails the processes that promote meaningful participation and awareness raising through information sharing, education and communication,
Legislation and regulation. This covers responsive regulation that provides the right incentives needed to improve upon enforcement management
Levels of Service. This deals with planning and designing appropriate services to suit the demand of different target groups
Research and Development. This relates to identifying and adapting appropriate technologies and modern techniques for change management
Monitoring and Evaluation. This involves appraising and assessing progress in policy implementation and the effectiveness of policy actions
Policy on Financing and Cost Recovery. The sector policy on financing and cost recovery is aimed at recovering operation and other costs on services whilst cushioning the poor. In fact cost recovery in most areas is targeted at the middle and high income population as for example, the free Communal Container service for solid waste at low income areas as against house-to-house services for those who can afford it.
Source: Ministry of Local Government and Rural Development (2008). The revised policy is currently awaiting Cabinet approval.

The Environmental Sanitation Policy 2008 stipulates that the Metropolitan, Municipal and District Assemblies take action to make by-laws to regulate environmental sanitation and prevent pollution within their areas of jurisdiction. The policy outlines the following four distinct functions to be carried out by the Assemblies with regard to environmental sanitation. These functions are the provision of waste management services, public health management services, environmental monitoring services as well as planning, monitoring and public relations. Policies tend to reflect the current thinking of solid waste management and provide a general assessment of the prevailing situation in the country and strategies (Table 2.2). Private sector involvement in solid waste became prominent in the 1990s and since then more PSI is seen in the solid waste sector. The policy states private sector provision of 80% of SWC. The evolving practices and roles are presented and discussed in chapter 5.

Table 2.2 Solid Waste Aspects of Environmental Sanitation Policy (1999, 2008)

Particular areas of policy	Policy directives and decision making
Solid waste management by Waste Management Department (WMD)	o All solid wastes generated in urban areas are regularly collected and disposed of in adequately controlled landfills or by other environmentally acceptable means by Waste Management Department (WMD), o At least 20% of the solid waste collection service is done by individual Assembly and 80% provided by the private sector.
Private sector involvement in SWC	o Involve the private sector in the provision of waste collection services, and supervision of the private sector by the WMD, o City is to be zone into service areas, private companies to be given monopoly in a zone with population less than 15000, o Private sector shall operate within the policies, regulations, supervisory and licensing arrangements set up by the public sector, o Full cost recovery where possible,
Environmental monitoring, and public health education	o Monitoring environmental health standards and sanitary regulations, o Educate the people on public health, environmental, sanitation issues,
Legislation, by-laws enforcement, and regulation,	o Promulgate and enforcement of the by–laws on sanitation together with national laws, o Strictly observing and enforcing environmental health standards and sanitary regulations, o By-laws are to be enforced by the Environmental Health and Management Department of the Assembly,

2.4.3   Formal rules for Regulating Private Sector Involvement

The regulation of the waste business and private sector involvement in solid waste collection is the mandate of the Ministry of Local Government.

The regulation seeks to promote competition through legal restrictions and regulatory rules and controls regarding market entry and exit, the capacities of companies operating in the waste market, the service standards, and user charges. The local governments are required to outsource solid waste collection to private sector service through contracts and also conduct regular monitoring of the quality of service rendered by the private service providers and sanction any non-compliance according to the regulations and terms of the contract.

The formal rules (policy, legal and regulation) and the contractual arrangements between the principal (local governments) and the agent (companies) are necessary local conditions for solid waste management. These rules include the Environmental Sanitation Policy, Local Government Act, Local Governments By-law, National Procurement Act and other national legislative instruments that provide rules for solid waste management. The National Procurement Act (2003) makes it mandatory for the local governments Tender Boards to use competitive bidding to select companies. The analysis of regulation by comparing the elements of the current regulatory practices against the regulatory rules and legislation is presented in chapter 8.

2.5   Conclusion

This chapter presented an overview of solid waste management and reviewed the institutional and regulatory context within which the private sector operates in Ghana. The Local Governments (Assemblies) have the mandate and legitimacy for solid waste management. The responsibilities of the local governments include (1) implementation of policies, (2) solid waste service provision, (3) by-law legislation, regulation, monitoring and enforcement of by-laws and solid waste management rules. The Ministry of Local Government oversees the implementation activities of the local governments.

The review shows that there are policy and legislative instruments for local service delivery including private sector involvement in solid waste management. The policy, legal and regulatory aspects of solid waste management are the external environment within which the private organisations operate. The actual use of the formal rules (external factors) and the influence on private sector performance will be explored in chapters 5 and 8. The next chapter discusses the theories and market mechanisms that shape private sector involvement in solid waste collection.

3 Theoretical Framework of Private Sector Involvement in Solid Waste Service Provision

3.1 Introduction

Chapter 1 provided the background, context of the study, research problem and objectives. An overview of solid waste management approach and the institutional arrangements (policy, legal and roles of institutions) for the solid waste sector in Ghana were provided in chapter 2. This chapter provides the reasons for private sector involvement in solid waste collection as well as the theoretical perspectives of market mechanisms and frameworks for understanding the factors explaining private sector performance in solid waste collection in the developed and developing countries.

This chapter is divided into seven sections. Section 2 reviews government and market failures in solid waste collection. Section 3 reviews management of public service and market mechanisms. Section 4 reviews the theoretical issues on the rationale for private provision of solid waste service. The context and theoretical arguments that have influenced the introduction of private sector in solid waste collection are reviewed. Section 5 reviews the theoretical debate on explanatory factors of private sector performance, including the private sector efficiency over public provision, and the debate concerning the service quality and efficiency gains of the private provision within the private sector. The section also reviews the explanatory factors of efficiency to identify the factors to be operationalised for this study. Section 6 identifies the enabling factors and drivers for improving performance to be used as lens to explore the case in Ghana in this study. This section presents the theoretical and conceptual frameworks used for analyzing the performance of the private sector companies and the explanatory factors of performance. The last section draws conclusions from the review on the performance explanatory factors to be explored.

3.2   Government and Market Failures in Public Service Delivery

This section reviews government intervention due to public good nature of solid waste service and the burden of solid waste collection on governments in developing countries. The reasons for government failures associated with the public services are also reviewed. The section further explains why there were market failures associated with the public services, and therefore necessitated the need for private sector involvement with the public sector playing a key role in facilitating, regulating, and monitoring the activities of the private sector.

3.2.1   Government failures in solid waste collection service

Government intervention has been defended by pointing to the existence of market failures –'public and merit goods and services', externalities, increasing returns to scale, and information asymmetries. There is considerable literature on situations where it is more appropriate to keep the provision of public service directly under public authority (Ascher, 1987; Gidman et al., 1999; Kessides, 1993; Roth, 1987; Cointreau-Levine, 1999). In a situation where externalities exist and where it is difficult to charge for service or exclude those who cannot pay, public provision is better than private (Ascher, 1987). The reason for government intervention in solid waste collection is the existence of externalities such as health hazards, unsightly litter across neighbourhoods, rodents, insects and smell within and across cities, which would come about if solid waste is not collected.

In developing countries, government intervention typically takes the form of providing solid waste collection services free of charge and raising general revenue through city taxes. But providing service at a zero price does not mean that people will take advantage of it because they must still make some effort to bring the waste to the point for collection (Walsh, 1995). From individual's point of view, throwing a litter on the ground is 'cheaper' than carrying it to the litter basket. An alternative solution to the problem of externalities is strict enforcement of existing laws which forbid littering and unsightly accumulation of solid waste. Walsh argues that since such laws are required even when the service is provided at a zero price, providing the service at a zero price may not significantly reduce the negative externalities. In short, the externalities associated with solid waste provide a strong argument for government provision of solid waste collection.

However, some argue that government failure in service provision has led to private sector involvement. There have been attacks on public sector 'direct provision' of public service. The critics contend and argue that without the discipline of competition, the public sector develops lazy habits and has no incentives to control costs or deliver effectively (World Bank, 1997, 1994). Batley (2001) argues that government fails to produce the optimum allocative efficiency (service match users preference and/or extent to which charges cover service cost) since non-competitive provision leads to allocative inefficiency and productive inefficiency. An allocation of resources is allocative inefficient if allocation does not match users preferences and/or are charged at prices excessive or subsidized and do not reflect providers real cost. Productive inefficiency occurs where resources are not used optimally to maximize outputs. Governments could not cope with the provision of public services, especially in the developing countries, due to inefficiency resulting from the non-competitive nature of service provisions, increasing financial burden of the public sector, and inadequate capacity of the public sector. Box 3.1 presents government failures in service provision (Grand, 1991). As a result of these government failures the private sector involvement (PSI) in service provision became

prominent in the 1970s in the developed countries and later on in 1990s it
began in the developing countries.

Box 3.1 Government failures in service provision

The argument of government failures is that public provision is associated with the fol-
lowing (Grand, 1991):
- Monopoly of public services, and markets are not 'contestable' (lack of competition)
- Absence of competition reduces incentives to keep cost to a minimum
- Allocative and productive inefficiency in service provision
- Low service coverage due to subsidies (demand for services exceed supply)
- Government failure to regulate service provision and public sector agents
- Budget maximization by the bureaucrats and self-interest bureaucrats

---

Private sector involvement has a lot to do with market and government
failures (van Dijk, 2008b). It is argued that private sector involvement is a
way to maintain market discipline and secure private finance for solid
waste collection and management. The rationale for PSI in solid waste service provision is that private sector brings management expertise, technical
skills and ability to finance investment and achieve greater efficiency and
lower cost (Cointreau, 1994; Bartone, 2001). The expected benefits of PSI
included providing access to private finance, better service provision (improve service quality), cost efficiencies and capitalizing on the strong
points of private sector (van Dijk, 2008b).

Greater efficiency in service provision is normally achieved through introduction of commercial principles and by paying attention to customer
satisfaction (World Bank, 1994). Even though public sector efficiency has
also improved in many cities in developed countries through the introduction of competition and commercial principles, private sector is seen to offer a means of achieving greater efficiency (Bartone, 2001).

However, for sustainable private sector provision and for these benefits
to be derived, there is the need for assurance of cost recovery charges and
enabling environment to recover investments. The degree of competition in
the solid waste market, the public-private relationships for the service pro-

vision and the market mechanisms are believed to provide enabling environment for successful PSI.

As a result of government failure in direct service provision, there is a shift in the role of the public sector from 'direct provision' of service to 'indirect provision'– formulating policies, setting service standards, financing, enabling, regulating, and monitoring the service providers. The public sector is playing a leading role of purchaser (buyer) on behalf of citizens through subsidies and user charges. The private sector is also taking on an increasing role as provider of public services (seller), and being regulated by the public sector through controlling charges and service quality. The private is regulated to correct market failures due to the 'public good' nature of solid waste service. The following section reviews the market failures in solid waste collection and why the need for regulation of private sector and the market.

3.2.2 Characteristics of Market Failures of Solid Waste Collection Service

Solid waste collection can be considered to have 'private good' attributes, and therefore it is possible to charge users at economic price and exclude not-payers or those who are not willing to pay for the collection service (Batley, 2001). However, those who are excluded will dispose of their solid waste on public or private land on individual basis. The indiscriminate waste disposal becomes an environmental health hazard or nuisance affecting society (negative externality), although the private households that generate waste may consider themselves to have done their duty by removing waste from their private domain.

Solid waste collection service can be argued to have 'public good' characteristic because such service has externality, non-excludability and non-rivalry problems and thus wholly private competitive market for their provision may not work (Cointreau-Levine, 1994; Awortwi, 2003). The high negative externality (pollution from uncollected waste) makes solid waste collection service a 'public good or service'. Public good or service is defined as goods which are inherently available to all and for which one person's use does not exclude or diminish their availability to another user. Externalities of uncollected solid waste do not encourage market provision by the private sector alone without government intervention.

Furthermore, some researchers have argued that sanitation and solid waste services will be under-supplied if markets alone are left to determine their provision (World Bank, 1994; Contreau-Levine, 1994). This is because if the environmental and social costs of the services are fully incorporated in user charges that fully recover costs, it may not always be affordable and private providers cannot be relied on to make investments. Box 3.2 sets out some of the main reasons why market provision may fail and government may have to intervene. It is argued that, goods or services characterized as 'public goods' are to be provided by the government and not through the market, because, the role of the government is to ensure the welfare of its citizens (Walsh, 1995). It is argued that markets fail to produce the optimum services and there will be the need for government interventions (Walsh, 1995). In other words, public goods and services cannot be adequately produced or provided by the private market.

Box 3.2 Nature of Public Goods and Services

The arguments of the market failures in literature include the following (Walsh, 1995):

- *Public goods and services* cannot be adequately produced or provided by the private market since they have *non-excludability* and *non-rivalry* problems. Excludability concept describes the situation where the operator of the service should be able to exclude potential users who are not willing to pay. If people cannot be excluded from the benefits of these services and are aware of the fact that it is impossible to exclude any member, they do not have an incentive to contribute to the provision of such goods and the service becomes a public good. But when a potential user can be excluded from benefiting from the service without paying the price, and no other alternative way of free-riding is available, the service becomes a private good. On the other hand, public goods and services may be described as rivalry or non-rivalry depending on the degree of diminishability once a user consumes a part of it. A service is non-rivalry when any resident can enjoy the benefit of the service without preventing the benefit of another user.
- *Equity or 'merit' goods and services* should be provided for society, and therefore

cannot be adequately provided by the private market. These are goods which are bene-
ficial for society that all may adequately partake in the service, but the individuals are
likely to underuse it.

- *Externalities* of the service do not encourage private provision. Externalities arise
when the production processes of an organization give rise to social cost (negative externalities as a result of pollution) or benefit (positive externalities) which is not considered in the pricing. Externalities occur when a cost or benefit that is borne by society is not reflected in the market price for the service or product. When producers do
not pay social cost of pollution, negative externalities, there is likely to be over-
production and over-consumption. On the other hand, when there are benefits that are
not taken into account, positive externalities, there is likely to be under-production
and under-consumption (Walsh, 1995).
- *Increasing returns to scale,* where there is continually decreasing unit cost as the scale
of production increases, encourages natural monopoly and monopoly power. Monopoly power is present when a single firm dominates a market as a sole service provider
and can have the capacity to exploit this position to the detriment of the customers.
Where there is monopoly power to provide services, customers would not demand because the firm has an incentive to charge prices significantly in excess of the marginal
costs of service. There is also lack of competition and incentives for innovation and
investment to provide goods and services at their lowest cost (productive efficiency).
- *Information asymmetries* between providers and users of the goods and services,
where users have too little knowledge to make informed choices or the public sector
have too little information on cost to enter into contract with private sector for provision of public services.

Increasing returns to scale (economics of scale) for solid waste collection is possible when a whole city is given to one company. Increasing returns to scale usually leads to natural monopoly which requires high capital
investment and encourages monopoly power of the provider. In solid waste
collection cities are sub-divided into districts or zones and given to a number of companies to have zonal monopolies over different service areas, to
foster competition. Even with the zonal monopoly, it still requires government intervention otherwise market will fail to provide the service for total
coverage of the zone.

In solid waste sector, there may be information asymmetries between
public and private sectors and/or between the service providers and the customers or users. Information asymmetries occur when the public sector and
customers do not have adequate information on which to base purchasing

decisions, and service providers may be able to exploit this situation when they have greater information. For solid waste collection as a visible service, customers may be able to judge service quality based on reliability and satisfaction factors, but may not know cost of service information.

Solid waste market will fail to perform efficiently due to these market failures (World Bank, 1997). Batley (2001) argue that the private sector alone will fail to provide the service through market due to certain situations. First, the private sector has no incentives to provide services for which it is not feasible to charge in proportion to use. This arises for 'public good' where users cannot be excluded. Second, private sector will not provide efficiently where market is non-competitive. Third, private sector on its own will ignore the costs and benefits of those who lie beyond the reach of the market and cannot afford. Finally, the market will not achieve socially acceptable levels of equity – merit goods. These are the reasons for government intervention to provide services to the citizens. Solid waste collection service is a public good, and therefore citizens expect the government agency responsible for waste management to act and keep the environment clean.

However, government intervention has also failed due to budget maximization by the bureaucrats and self-interest bureaucrats resulting in low service coverage and inefficient service provision. Therefore, government and market failures led to the need for regulated private sector involvement. Besides government and market failures, the new role of public sector in the management of public service and engagement of private sector is reviewed in the next section.

3.3   Market Mechanisms for Public Service Provision

Public service is defined as any service available to the public, whether provided publicly or privately (Roth, 1987). The provision and delivery of public services are the responsibility of government. As part of the public sector reforms in the delivery of urban services, governments and development agencies began, in the past two decades, to vigorously promote the private sector as a provider of urban solid waste management services, especially in the developing countries (Cointreau-Levine and Coad, 2000), to improve service efficiency[9] and effectiveness[10] in solid waste management. The use of market-type mechanisms (outsourcing of public services to the private) and the enabling role of central/local governments are key elements of the new public management (Aucoin, 1990; Hood, 1991; Helmsing, 2002, Batley and Larbi, 2004). The following section discusses the conceptual issues of the new public management, decentralization of management of public services, and the enabling role of the government.

3.3.1 New Public Management and Managerialism

Management of Public Service

The new public management (NPM) is a set of reforms meant to shift the public sector management from bureaucratic to post-bureaucratic organizational arrangements (Hood, 1991). In the quest for efficiency and effectiveness in government, the adoption of private sector management techniques and practices was advocated to deal with the problems of the old public administration by replacing hierarchy with markets in the hope this would minimize the dysfunctions of traditional public administration. There is an emphasis on management and markets instead of administration and hierarchy respectively. For example, efficiency is considered more important than adherence to rules; and in the same way effectiveness and results take precedence over legality; flexibility and adaptation are more vital than predictability and responsibility; and competition with private sector delivery is preferred to delivery by government monopoly and bureaucracy (Batley and Larbi, 2004).

---

[9] *Service efficiency* means a service is efficient if the available resources are used in the best possible way.
[10] *Service effectiveness* means the extent to which the objectives of the service have been met in practice.

The new public management theory and practice have evolved since the term was first used by Christopher Hood in his article 'A Public Management for all Seasons?' published in 1991. Hood's original conception of NPM and key elements originate from most of the structural, organizational and managerial changes in the public services of the Organisation for Economic Cooperation and Development (OECD) countries and dated back to the late 1970s through the 1980s (Dunleavy and Hood, 1994:9). NPM can thus be defined as a set of particular management approaches and techniques, borrowed mainly from the private for profit sector and applied in the public sector. It is sometimes perceived as an ideology based on the belief in the efficacy of markets and competition, and in business-like management ideas and practices (Hood, 1991, 1995).

NPM's key elements may be categorized into two main strands; on one hand are ideas that emphasise *managerialism* – managerial improvement, and on another hand are techniques and practices that emphasise *markets and competition*. The first strand of NPM is managerialism and is defined by Pollitt (1990) as involving continuous increase in efficiency, use of technology, enhanced productivity, and more managerial role and managers being given the right to manage. The managerialism in this context includes restructuring by decentralizing the management of public services to emphasize managerial autonomy and professionalism and it involves disaggregation, downsizing and emphasis organizational and individual performance management (Batley and Larbi, 2004). The aim of managerialism is to gain more effective control of work practices.

The managerialism aspect of the new public management is more about how to improve the management of activities that remain under public sector ownership by applying private sector management practices. In the quest to re-organise public service delivery, there is increasing emphasis on output controls in resource allocation and a shift in the form of control and accountability mechanisms from inputs and processes to performance targets and output objectives at both individual employee and organizational levels in the public sector (Batley and Larbi, 2004). A logical development from the emphasis on measuring and rewarding performance is that audit institutions now undertake performance auditing in addition to the traditional compliance auditing. A number of countries have also introduced performance agreements or contracts between public enterprises and government and between funders and providers of services (Batley and Larbi, 2004).

The second strand of NPM is based on indirect control and is characterized by continual improvements in quality, emphasis upon devolution and delegation to the private sector, emphasis on contracts and markets, measuring performance, increased emphasis on audit, inspection and information systems (Pollitt, 1990). The second strand emphasise markets and competition as a way of giving 'voice' and 'choice' to users and promoting efficiency in service delivery (Batley and Larbi, 2004). *This strand is also called 'market mechanisms' and includes contracting out and other forms of private sector participation, creating synergy between public and private sectors, introducing user charges for public services, and emphasizing quality and responsiveness to users.* Two main arguments are given for why such market mechanisms (i.e. both commercial principles and competition) will prove successful. First, this will reduce the financial burden on the state. Second, it is argued that reforming service provision along market mechanism lines will improve quality of service. Improvements in productivity and pricing would permit more effective delivery of service in response to demand (World Bank, 1994). This study attempts to use the second strand and managed market theory to explain the involvement and regulation of the private sector. The theoretical case is that the involvement of private sector in service provision will lead to improved pricing and user charging, better service quality, improved productivity, and responsiveness to users.

It can be noted that not all the administrative and management reforms taking place in the public sector fall under the rubric of new public management. For developing countries, non-NPM reforms include various efforts at capacity building, political decentralization or devolution reforms (Batley and Larbi, 2004). The extent of implementation of NPM in most countries is at different stages of reshaping government roles, restructuring public organizations to enable greater managerial autonomy, redesigning and improving service delivery to be more responsive and efficient, redesigning systems of performance and accountability, and revitalizing human resource capacity and organizational performance (Batley and Larbi, 2004).

Elements of New Public Management

The reforms in public service delivery to improve governance brought about the introduction of the elements of the New Public Management (NPM) theory and it is intended to result in more effectiveness, more efficiency and financial sustainability. The NPM theory helps to guide reform process because of the combination of useful principles for reform and the possibility to translate those principles into effective instruments (van Dijk, 2004). The NPM formalizes the link between the service provider and its environment by using contracts. There are four elements or principles of the NPM: autonomy of organization and decentralization of decision making process, accountability for results, market orientation, and customer orientation (van Dijk, 2006). Van Dijk (2004) argue that with increased financial autonomy of the organization, a mechanism is triggered which is likely to lead to: a) increased customer orientation as the service provider needs to generate its income from the customers, b) increased managerial autonomy, c) increased accountability to local institutions, d) a drive for increased efficiency as the pressure to reduce costs increases.

The increased autonomy and decentralization of authority has external and internal components. The external component refers to decentralizing authority to autonomous organizations, which become responsible for providing services. The internal component relates to decentralizing authority within the organization. The autonomy of organization in the sense of having discretional powers and financial autonomy allows the organization to perform. The increased financial autonomy leads to a drive for increased customer-orientation. By cutting out subsidies and price distortions and increasing cost recovery, decision-making is facilitated. Financial autonomy implies the organisation looks for different sources of finance and this does not have to come from the government, but could also be provided by donors or result from direct access to capital markets.

The element of accountability for results has an external and an internal component. The external component concerns accountability for results produced by the organization as a whole. The internal component relates to accountability for results achieved by individual managers and employees within the organization. Accountability for results within the organization as well as external is through the use of contracts with performance indicators, monitoring, and reporting of results. The charging for service makes the organization become more accountable to customers and other organizations. Accountability requires activities that ensure external pressure. Consumer agencies and the press act as watchdogs over the organization's performance, but often the customers are not sufficiently organized to influence the reform agenda. Stimulating the development of stronger customer agencies could be a way of increasing customer influence on the reform process and service quality.

Market orientation means a greater use of markets and introduction of market-style incentives such as contracting-out and charging for the service. It involves a range of reforms starting with user charging for full cost recovery. After gaining experience with internal contracting and benchmarking, these instruments can also be used to involve the private sector, or to compare with other (private) operators.

The improvement in customer-orientation leads to increased fee collection, which in turn allows the organization to spend more on investments from its own resources (van Dijk, 2004). Customer orientation helps to find out what customers really want and what their problems are. It is important that the service provider reacts promptly and adequately to the complaints of customers. Services are performed better due the visibility of the service to the public and local institutional factors that compel the provider to be responsive to users (Batley, 2001).

3.3.2   Markets for Public Service Provision

Market Theory

The Public Choice Theorists argue that public delivery by public officials is not efficient due to the rigidness and more formalized rule system (Down, 1967), budget maximization by bureaus (Niskanan, 1994) and the uncontrollability of the bureaus (Tullock, 1965).The public provision unit was unresponsive to users' choice and inefficient, and therefore private provision was seen as an alternative for cost efficiency and better service quality. These theorists argue that contract market and competition lead to efficiency gains.

By definition, market is an exchange mechanism (transaction) involving buyers and sellers (market participants) in a market place where services or products are involved in exchange. Market as a process involves the exchange transactions of buyers and sellers and the actions of other bodies that facilitate such transactions. Ulkuniemi (2003) identified elements of the market from the buyer's perspective to be the buyer's own needs, the industry demand, industry supply, object of exchange, exchange mechanism, and exchange counterpart.

The neo-Liberal free market theory does not work for public goods and services. The market theory indicates that effective market requires large number of buyers and sellers who are well informed about the quality, price and availability of goods and services being exchanged (Brown and Potoski, 2004). The 'consumer markets', where there is competition in the market, work for private goods and services. Market fails when there are too few buyers and sellers or when information problem prevents transferable property rights.

The managed 'producer market' for public service seeks to address market failure (reduce information asymmetry) and stimulate market competition. A well functioning solid waste market requires stimulation of competition for the market, reduction of information asymmetry, and reduction of transaction cost. Local government and households (buyers) engage companies (sellers) in the 'producer market' to supply solid waste collection service in a market which has certain characteristics (sole buyer, regulated entry and exit, regulated market prices). For public goods and services, there is competition for the 'producer market' where public and private firms (sellers) compete to supply goods and services to government which remains the main and sole direct purchaser (buyer) (Hoggett, 1996).

In such a producer market the government has power to dictate many of the rules of the game and control local government contracts to the interest of politicians (Hoggett, 1996). The rule of the market and the positional powers of the players (prices, standards) are determined by government. There are three market control strategies for public service markets: creation of decentralised units with centralised control over strategy and policy; managed competition through contracting to public and private sectors at local government level; and performance management and monitoring (Hoggett, 1996). The management of market-based public service involves separation of purchasers and providers, use of contracts and commercial operations based on prices and user choice (Deaken and Walsh, 1996). The performance and quality management are central to the market-based management of the public service.

Market mechanisms

Besides the internal re-organisation with managerial emphasis in public sector, a number of countries are using market-type mechanisms (OECD, 2000; Walsh, 1995). The market-type mechanisms are in several forms – increasing focus on the customer, user fees or charges, contracting out of service delivery to the private and voluntary sectors, public-private partnerships. The market-based instruments seek to correct market failures and to allocate resources efficiently.

The market mechanisms of contracting out service, user fee charging and customer orientation are regulated in a 'producer market' to ensure efficiency gains and better service quality. Contracting out has become one of the key forms of public service delivery. This and other forms of private sector participation are being used in public service that cannot be privatized (outright sale of assets) for strategic, political or other reasons. Contracting out is increasingly being applied as a management tool, especially where a complete market solution or outright privatization cannot be employed. As a market-type mechanism, it is used to foster competition for the market with governments contracting with private sector organizations which competitively bid to provide services (Walsh, 1995). Contracting out thus puts competitive market forces directly at the service of government. Though contracting out or outsourcing of public service delivery is not new, what is new is the extension of the practice to cover activities that have traditionally been carried out by in-house public delivery arrangements. Where output is easily specified but direct competition is impossible, competition is managed through various forms of contracts.

Charging for services is seen as a mechanism for cost recovery of solid waste services, and also as a way of addressing issues of inefficiency and ineffectiveness in the delivery of services. The rationales for the introduction of user charges for public service and withdrawal of subsidies are the efficiency, equity, quality arguments and the need for diversity in the financing of public service to reshape public spending (Walsh, 1995; Batley and Larbi, 2004). The lack of efficient prices for public service can be seen as leading to unintended subsidy, where the better-off are subsidised. The arguments for pricing and charging are clearly strong. The opponents of user charging argue that the introduction of charges for solid waste will lead to negative externalities and therefore should be provided for free (Walsh, 1995).

This study will deal with market-type mechanisms such as increasing focus on the customer and charging for service (households' involvement), as well as the private sector involvement through public-private partnerships arrangements for contracting out of service delivery to the private sector.

Market Organizations for Public Service Provision

The solid waste management has four fragmented stages – storage, solid waste collection and transport, treatment and final disposal – which can be organized under different forms of organizational arrangements. The deliv-

ery of the services may be organized in four different organizational arrangements depending on whether the service is public, private or community managed and the ownership of the assets (World Bank, 1994).

First, services are organised under direct public management which involves public ownership of assets and operated and managed by government organisation or department in accordance with politically determined
rules and procedures. The direct public mode describes service provision
organization which is publicly owned and managed. This option is in fact
the application of some basic private sector management principles to the
way public services have traditionally operated. There is application of
three principles: public operation by quasi-independent public department,
explicit contracts between government and public managers and a pricing
strategy designed to ensure cost recovery. For services which are 'public
goods' the providers in many cases are in the public sector.

Second, services are organised under delegated private management
mode which involves public ownership with operation and management
contracted out to the private sector organization. This describes provision
of services through extensive out-contracting of management tasks to a
private company – in a form of private sector participation where the public sector owned the assets. The private firms are the operators whose aim
is to make profit. There are six variants of delegation to the private sector,
namely service contract, franchise contract, management contract, affer-
mage contract, build-operate-transfer, concession. These contracts will be
discussed in the next chapter.

Third, services are organised under direct private management which
involves private ownership of assets and management through economic
markets, in which a variety of private enterprises operate for profit. The
goal of this organizational mode – private ownership and operation – is to
reduce monopolies and use market mechanisms. By isolating the natural
monopoly segments of an industry, unbundling promotes new entry and
competition in segments that are potentially competitive. Failure to un-
bundle can constrain an entire sector to monopoly provision even when the
numerous activities can be undertaken competitively (World Bank,
1994:53). Finally, services are organised under community and user groups

involving community based organisations (CBOs) and self-help groups (World Bank, 1994; Gidman, et al., 1999). The community and user groups in most cases involve informal actors in solid waste management.

There is a mixture of market suppliers for solid waste collection in Europe and USA. These are private firms, public owned firms operating under commercial law, mixed public-private firms (joint-venture) and public units operating by bureaus under statutes. The debate on market framework (use of mixture of market suppliers) for local service delivery is which markets deliver cost efficient and better quality service (Boyne, 1998).

3.4   Theoretical Issues on Private Sector Involvement

This section reviews the forms of Private Sector Involvement (PSI) and the reasons why PSI provides option for addressing the problems of government and market failures. The section also reviews the debate on private sector performance over public sector, as well as the factors influencing service quality and cost recovery in developing countries.

3.4.1   Forms of Private Sector Involvement in Solid Waste Service

Emerging PSI and PPP

van Dijk (2008a) defined Private Sector Involvement (PSI) as participation ranging from complete private provision of public services to complete public provision with minimal private sector activity. PSI is a generic term describing the relationship formed between the private sector and public bodies often with the aim of introducing private sector resources and/or expertise in order to help provide and deliver public sector assets and services (Schouten, 2009). The central element in a PSI is the (often formal) arrangement or partnership that the actors conclude with one another. In a PSI the relation between the private provider and the public buyer is central, and in many cases such relationship or arrangement is shaped in the form of a contract (Van Dijk, 2008b).

Public Private Partnership (PPP) can be described as a contractual relationship between public sector agency and private sector where resources and risk are shared for the purpose of delivering a public service or providing infrastructure. PPP in developing countries has been slow especially in the Sub-Saharan Africa countries, although there is increasing private sector involvement (PSI) uptake in French-Speaking Africa (Li and Akintoye, 2003). PSI can vary from the provision a service to outright ownership of facilities depending on government control and private economic scale. Li and Akintoye (2003:9) drawn from analysis that there are five forms of PSI: service contract, leasing, joint venture, concession and privatization. Box 3.3 further gives explanations on the forms of PSI in public-private arrangements.

The simplest form of PSI is service contract, and in this form of PSI, the public sector retains control over policy decisions. The strongest form is joint venture where co-responsibility, co-ownership for service delivery, shared resources and shared returns are the core of the partnership. Grant (1996) points out that the core of partnership in PSI involves shared authority and responsibility, joint investment, sharing of liability/risk taking, and mutual benefit. PPP is characterised by five features: (1) involves two or more partners, (2) each partners is principal, but in situation of power relation some partners are agent, (3) involves enduring and stable relationship among partners, (4) involves transfer of resources (money, authority, expertise) and risk, and (5) mutual sharing of responsibility for outcomes or activities and benefits (Li and Akintoye, 2003).

Box 3.3 Forms of PSI in Solid Waste Collection

Forms of PSI or methods of private sector participation in literature include the following:

Service Contracting. Government awards a finite-term service contract to a private firm to provide solid waste services, and government pays the firm for the services delivered. Alternatively, government awards a management contract to a private firm to provide management oversight of others who are providing solid waste services. Service contracts transfer to private providers the responsibility for delivering a specific service at lower costs or obtaining specific skills or expertise lacking in the public sector. Another form is *management contract.* Management contract involves transfer of responsibility to private provider for managing an operation such as supervising collection operations, sanitary landfill or compost facility. It increases the autonomy of management and reduces the risks of political interference in the day-to-day operations of public entity.

Franchising/Asset leasing. Government grants a private firm an exclusive monopoly to provide a specific type of solid waste service within a specific zone. The firm collects its own revenues from generators within the zone or from the sale of solid waste by-products removed from the zone. The model uses competition between bidders for the franchise as a substitute for competition between providers in the market. In this way franchising mimics the effect of market competition. Franchisees pay rental fee to public sector. *Private subscription:* Government licenses private firms to compete with each other in providing solid waste management services. No firm has a monopoly within a zone and price regulation is not required. Each firm collects its own revenues from its customers or "subscribers". (Also known as "open competition".)

Joint Venture. This involves co-responsibility and co-ownership for service delivery. This is 'true' PPP where there is shared resources and shared return.

Concession. Government allows the private sector to utilize one of its resources, in this case solid waste, for profit-making purposes. Concessions typically involve construction of major long-term facilities to sort, treat transfer, or dispose of solid waste. Government may pay a tipping fee or service charge to defray part of the costs of processing the solid waste, but sale of the concession's product (such as compost), or service fees paid by non-government customers typically cover the remaining costs. Government provides a guarantee of flow control, so that amounts of wastes received closely match facility design capacity. Most concessions are operated on a "take or pay" basis, where tipping fees are paid even if the guaranteed daily quantity of waste is not provided.

Privatisation. This involves sale of state-owned asset to private sector.

Source: (Cointreau-Levine and Coad, 2000; Li and Akintoye, 2003; Batley and Larbi, 2003):

PSI with Investment Risk

Figure 3.1 presents the extent of risk and forms of PSI involvement according to classification based on investment risk in solid waste collection in order of increasing private sector risk. The element of risk sharing in investments is the defining characteristic used to distinguish public-private partnerships from other arrangements of PSI (van Dijk, 2008a).

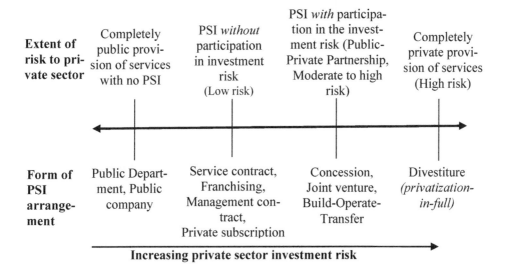

Extent of risk to private sector	Completely public provision of services with no PSI	PSI *without* participation in investment risk (Low risk)	PSI *with* participation in the investment risk (Public-Private Partnership, Moderate to high risk)	Completely private provision of services (High risk)
Form of PSI arrangement	Public Department, Public company	Service contract, Franchising, Management contract, Private subscription	Concession, Joint venture, Build-Operate-Transfer	Divestiture *(privatization-in-full)*

Increasing private sector investment risk

Figure 3.1 Extent and Forms of PSI for public services

Source: van Dijk (2008a): Private sector involvement in service provision.

The different forms of PSI without participation in investment risk have certain risks, but they are not investment risks (van Dijk, 2008a; Hardcastle and Boothrouyd, 2003). For example, in solid waste collection private companies invest in new fleet of vehicles but the contract duration allows for the depreciation of the vehicles to be recovered. PSI with participation in investment risk (also called public-private partnerships (PPP)) involves an establishment of partnership which can be achieved with or without any competition or shift in ownership (Schouten, 2009). The forms of PSI are public-private arrangements for public service provision and concern contractual arrangements between local governments (public sector) and the private sector companies. These public-private arrangements involve transfer of responsibilities in terms of investment capital, management and risks.

Risk in PPP is view differently by parties involved and the importance of risk to parties varied (Hardcastle and Boothroyd, 2003). This is because some risk issues are particular to a specific party. A theoretical review of risk by Hardcastle and Boothrouyd (2003) reveals a vast range of risk issues, but seven key among them are:

- o *Investment risk* – it involves loan repayment risk, default risk, and the inability to recovery funding cost,
- o *Legal risk* – it involves new legislation and regulation after transaction that affect the partnership,
- o *Market risk* – it involves change in demand, interest rates, exchange rates, energy prices, as these risks tend to reduce value of investment,
- o *Counterparty risk* – it involves non-adherence to contract obligations after contract is signed,
- o *Volume risk* – it involves lack of demand for the service may pose risk to the investor,
- o *Operational risk* – it involves system failure, human error, poor quality service,
- o *Increase cost risk* – it involves increased cost to maintain performance which may be higher than projected,

The private sector perspectives on investment in solid waste collection and management, especially in developing countries, are on these risks and the risk of recovery of investment. The management of these risks requires risk analysis (identification and classification of risk), risk response (allocation), risk outcome monitoring and control, and risk evaluation (Hardcastle and Boothroyd, 2003).

There are different forms of private sector involvement (PSI) in water, sanitation and solid waste collection around the world (van Dijk, 2004). Under different forms of inter-organizational arrangements for service de-

livery, government and the private sector retain different degrees of responsibilities in service provision, ownership of assets, financing investment, and operations management. There is now recognition that there is a
smaller role for government in the direct provision of services and a bigger
role for government in policy development, coordination and regulation.
From the beginning of the 1990s, reforms began to affect the inter-
organizational arrangements for service delivery – for the transfer of the
delivery of social and infrastructural services to the private and community
or NGO sectors (Odle, 1993; Cook and Kirkpatrick, 1995; Batley, 1996).

3.4.2 Rationale for Private Sector Involvement in Public Services

Economic theory suggests that when services are organized through market, competition requires numerous buyers and sellers such that no single
entity can dominate. Local Governments can influence competition
through regulatory, facilitative and monitoring roles. It is expected that
PSIs will provide opportunities for users to participate not only in paying
for the cost of service delivery, but also in decisions regarding the kind of
systems, technology, and the standard and quality of services that will be
delivered to them.

In theory, the involvement of the private sector in public services is to
improve the performance of public services and to attract private capital
for major investments (Rothenberger, Frei and Brugger, 2005). The PSI is
expected to foster high efficiency, minimize cost and provide better service
delivery. The rationale for introducing different inter-organizational arrangements for delivering solid waste services was to reduce cost of services. It is argued in the literature that contracting out leads to cost savings
and better value for money by removing the production of such services
from inefficient public bureaucracies that are more intent on satisfying the
wishes of producer groups than of consumers (Awortwi, 2004). Contracting out reduces costs in three principal ways. First, private companies are
free from Civil Service requirements, they use an incentive pay system and
have greater freedom to hire and fire workers and they employ part-time
workers, have less absenteeism and therefore are more flexible than the

public sector in reducing costs. Second, they pay lower wages than government agencies. Finally, private companies tend to pay their workers substantially lower fringe benefits (Helmsing, 1997:72).

There are four arguments that if services are provided by the private sector, governments will be able to: reduce subsidies and cost; reduce public borrowing by encouraging the private financing of capital expenditure on infrastructure; increase tax revenue from private operators; and reduce the public debt by selling assets (Rakodi, 2003:3). Furthermore, it is suggested that private provision will result in improved equity, since charging everyone for services will generate sufficient revenue to expand services to unserved residents and areas.

The new public management theory, principal-agent theories and public choice theory provide the right fundamentals for public-private partnerships in public services. The public choice theory indicates that the introduction of competition in awarding contracts is the critical factor in lowering costs and improving efficiency. There are substantial research findings on efficiency gains stemming from competition (Domberger and Hall, 1996; Domberger et al., 1986; Prager, 1994). Literature indicates that competition determines service efficiency and effectiveness (Awortwi, 2004).

Despite the reasons for PSI, there are three performance and contractual issues concerning greater private sector involvement in public service provision. First, there are problems with markets structure for the service delivery – diffuse or concentrated structure of organizations – and how to construct benchmark measures to monitor performance of the private sector (Walsh, 1995; Rothenberger, Frei and Brugger, 2005).

Second, the private sector management is likely to challenge ways of organizing public sector services, especially equity and coverage objectives of the public sector may not be met, if there are no incentives to serve the poor areas. Third, the nature of partnership contracts – contracts specifications, unclear definitions of tasks and responsibilities, and trust – may raise questions about outcomes or performance gains of the new arrangement. Finally, the relation of power between partners is likely to lead to opportunistic behaviour due to information asymmetries, lack of political commitment and can pose major problems to the contracting partners (Grimshaw, Vincent and Willmott, 2002; Rothenberger, Frei and Brugger, 2005).

The purpose of private sector involvement in solid waste service was to improve efficiency and effectiveness in order to protect public health. However, in developing countries, there are problems with arrangements for working towards public interest, policy objectives, transparency and coordination of public services, and how to achieve greater efficiency of PSI by contracting, monitoring and regulating. In public-private relationships, the danger is that the efficiency gains from contracting the private sector may be outweighed by the additional transaction costs of doing business between multiple actors – setting policy frameworks, coordinating, contracting and monitoring are done at a cost.

A greater efficiency gain can only be true if the public and private partners have the capacity to perform their new roles. Contract specifications and terms that establish duties and obligations, service quality, performance monitoring, rules for changing prices and sanctions hold the key to contractual exchange. For private agents to be efficient and effective to improve the quality of service delivery, they need to have sufficient capacity to contract, draw up business plans and implement them, and more importantly deliver the services to the satisfaction of Local Governments and the general public. The inability of private agents to meet many of these capacity requirements was confirmed to have ramification for poor quality of service delivery (Awortwi, 2003). The inability of the private agents in developing countries may be due to internal organizational factors, inter-organisational arrangement problems and external environment factors (policies, legal and regulation).

3.4.3 Public-Private-Community Partnerships and Households Involvement

Public-Private-Community Partnership (PPCP) is another form of partnership among public, private and community where the partnership arrangements cover a variety of formal or informal arrangements (Karanja, 2005,

Mwangi, 2003). These arrangements range from legal (guided by formal rules and regulations) to informal forms of cooperation and co-ordination. The partnerships involve the three actors (public sector, private sector, communities) working together for mutual benefits. A partnership is an enduring relationship between these actors (based on a written or verbal agreement, formal or informal in nature). The most successful and enduring partnerships are those that bring real benefits to all actors involved (Karanja, 2005). The principal aim of partnerships especially public and private sector is usually to improve service efficiency, effectiveness, share risks, responsibilities and financial burden. Under the partnerships, the private sector is expected to inject resources such as easy access to finance, knowledge and technologies, and also its dynamism and managerial efficiency.

Community development theory suggests the involvement of all stakeholders in development process helps to achieve an optimal solution. Those who received the service (customers) have a role to play in the service delivery, such as paying for the service and cooperation with service arrangements. Customer orientation is organizing and directing the services of the provider towards customers (Cullivan et al., 1988). The provision of efficient, effective, and equitable service is directed towards the customer. Being aware of this, customer orientation helps govern important daily operational decisions and actions. Effective service providers have workable means or mechanisms for customer interaction. These include emergency outlets or phone line, identified places where disputes about bills or service can be arbitrated, ways that interested customers can make suggestions in overall policy. Creative and cost-effective ways are sought to inform and educate the public.

Community involvement is contingent on the existence of some type of mechanism for discussion of common issues, for the resolution of differences in opinions, for the efficient and effective monitoring of service. This mechanism may be provided by local government, or may be provided by the service provider. Community involvement influences people's willingness to participate and increases participation and the probability of success of a development project or process. Community involvement allows the community to express their opinions and needs. High levels of trust, coupled with increased participation may also allow for greater consensus on the most important opinions and needs of the beneficiary. This increased ability to articulate needs and opinions will improve the ability of the beneficiary (community) to express their needs, which will help the provider to make better decisions in the service delivery or implementation and management of the development project.

3.4.4   Evidence of Private Sector Efficiency and Service Quality

Empirical Evidence of Cost Efficiency of Private Sector in Developed Countries

There are two kinds of studies on whether PSI is associated with private sector efficiency gains. One group of studies focused on econometric empirical and statistical analyses whereas the other group focused on country case studies. The econometric studies generally explain whether PSI (measured in terms of ownership or organization form) has positive impact on cost efficiency of the private sector. However, results of review of these cross-country studies (mostly developed countries) on water and sanitation and solid waste collection have been inconclusive (Prasad, 2006; Bel and Warner, 2008). On the other hand, the case studies in water and sanitation demonstrate that there have been some improvements (especially in productivity and profit), but the process is much more complex and the benefits are not automatic (Prasad, 2006).

The econometric empirical and statistical studies in developed countries on private sector costs efficiency in solid waste collection have been reviewed to find out more about the inconsistent results of the studies and to identify theoretically based reasons and explanatory factors why cost efficiency (cost savings) of the private sector are not consistently and systematically found. Table 3.1 presents summaries of econometric empirical studies on private sector cost efficiency in waste collection in the developed countries from 1960 to 2005.

Table 3.1 Solid waste collection cost efficiency in developed countries

Reference	Location	Year	Sample	Cost of private provision over public
Hirsch (1965)	USA	1960	24	No difference
Pier et al., (1974)	USA	1970s	22	No difference
Kitchen, (1976)	Canada	1970s	48	Public higher than private
Kemper and Quigley, (1976),	USA	1970s	128	Private more expensive
Collins and Downes, (1977)	USA	1970s	53	Private more expensive
Pommerehne and Prey (1977)	Switzerland	1970	103	Public higher than private
Stevens, (1978)	USA	1974	340	Private more expensive
Tickner and McDavid, (1986)	Canada	1981	132	Public higher than private
Domberger et al. (1986)	UK	1985	305	No difference, *with competition*
Dubin and Navarro (1988)	USA	1974	261	Private more expensive
Szymanski and Wilkins (1993)	UK	1988	335	No difference, *with competition*
Szymanski (1996)	UK	1994	>300	Private costs are lower, *with competition*
Reeves and Barrow, (2000).	Ireland	1995	48	Public higher than private
Callan and Thomas (2001)	USA	1997	110	No difference, *with organization form*
Dijkgraaf and Gradus (2003)	Holland	1997	85	No difference, *with competition*
Ohlsson (2003)	Sweden	1989	115	Private costs are higher
Bel and Costas (2006)	Spain	2000	186	No difference with form
Dijkgraaf and Gradus (2007).	Holland	2005	491	Private costs lower but erode with time

Source: Modified from Bel and Warner (2008)

Hirsch (1965) conducted the first econometric study of solid waste collection using data from 24 municipalities in USA (St. Louis County, Missouri). Hirsch found no significant difference in costs by municipal or private contract arrangement. His cost model provided an example that has been followed by many studies (Bel and Warner, 2008). His cost model controlled for quantity of solid waste, service quality, level of service (conditions) that affect input requirements, technology, population density, and form of financing arrangements (user fee or general budget).

Similar results have been found in other studies of USA municipalities, in Montana (Pier et al., 1974) and Missouri (Collins and Downes, 1977). They found no difference between public and private solid waste collection. Another study in USA by Stevens (1978) found no difference in costs in municipalities with population less than 50,000, but private monopolies were less costly in cities with over 50,000. This shows the importance of economics of scale for solid waste collection. A Connecticut study (Kemper and Quigley, 1976), found private solid waste collection had lower costs, but they did not control for heavier public solid waste collection in cities.

In the USA, Callan and Thomas (2001) found no difference in cost in solid waste collection. Bel and Costas (2006) study of 186 Spanish cities also found no difference in cost. Ohlsson's (2003) study of 115 Swedish cities found private provision more costly than public because of higher input and capital costs for private firms. Dijkgraaf and Gradus (2003) show no difference between public and private solid waste collection costs under competitive contracting among cities in Holland. Dijkgraaf and Gradus (2007) found no cost differences between cities that had started PSI earlier and that of public provision.

In summary, majority of these studies find no difference between public provision and private provision. While 6 out of 18 studies find cost savings with PSI, these results are not consistent over time – private sector performance has not been consistent. Only 3 studies found consistent lower costs with private provision. These include two works from Canada in the 1970s (Kitchen, 1976) and 1980s (Tickner and McDavid, 1986) and one more recent study in Ireland in the mid-1990s (Reeves and Barrow, 2000). Most recent studies, since 2000, on waste collection found no differences in costs.

Cost Efficiency and Service Quality of Private Sector Service Provision in Developing Countries

There are few econometric empirical and statistical analyses on private sector cost efficiency in developing countries. Studies on PSI in developing countries are mostly case studies with no econometric or statistical analyses of private sector cost efficiency and service quality. Table 3.2(a) summarises case studies on PSI in solid waste collection and management in developing countries. The recent case studies of PSI in some developing countries – for example, in Kenya (Karanja, 2002), in Ghana (Obiri-Opareh, 2002; Awortwi, 2003), in Tanzania (Kassim, 2006) and in India (Post et al, 2003) – showed that there has been increased coverage in some of the countries, but the service quality, efficiency and sustainability of private sector service delivery are still issues that require detail studies to identify explanatory factors for private sector performance and drivers for sustainable PSI.

Table 3.2 (a) Solid waste collection cost efficiency and service quality of service provision in developing countries

Reference	Case study	year	Number of cities	Service quality and cost recovery issues
Bartone (1991)	Latin America	1990	5	Private costs lower, quality equal/better
Post et al. (2003)	India, Ghana	2003	2	Quality improved, but cost higher
Massoud et al. (2003)	Labanon	2003	1	Quality improved, but cost higher
Karanja (2002)	Kenya	2002	1	Quality improved, but low coverage
Obiri-Opareh (2002)	Ghana	2002	1	Quality improved, partial cost recovery
Awortwi (2003),	Ghana	2003	3	Quality improved, partial cost recovery
Kassim (2006)	Tanzania	2005	1	Quality improved, no cost recovery

Source: author's review

Table 3.2(b) summarises three case studies by Karanja (2003), Kassim (2006) and Gafishi (2009) on PSI in solid waste collection and management in Africa. Karanja (2003) points to the fact that the most outstanding impediment to effective involvement of the private sector in solid waste management in Nairobi is the lack of regulation. The issues identified by Karanja (2003) and Kassim (2006) include weak regulation, low capacity of private sector organisations, low cost recovery, and poor service quality.

The private sector still faces challenging issues of inefficiency and low service quality due to some factors in the partnership arrangements and how company organizations are run, which this study seeks to identify. The rationale for the private sector involvement (PSI) in solid waste collection is to improve efficiency (reduce cost) and effectiveness of service delivery (service quality) through competition for the market and to ensure that the environmental aspect of sustainable development is integrated in solid waste management. But this, in most cases, is not realised in developing countries due to certain internal and external factors.

Despite the fact that the private sector may be cost efficient compared to the public sector, key issues to which will the focus of this study are:

- o Regulation,
- o Service quality
- o Private sector capacity.
- o Cost recovery,
- o By-law enforcement,
- o Households' involvement,

These issues have come up strongly in the past studies in Africa. These issues will be explored further in this study to find out the factors which influence the low cost recovery levels and the resulting poor service quality. The understanding of the internal and external factors related to these issues will help provide the way forward for interventions.

Table 3.2 (b) Solid waste collection cost efficiency and service quality of service provision in developing countries

Issues	Kenya (Karanja, 2003)	Tanzania (Kassim, 2006)	Rwanda (Gafishi, 2009)
Regulation of private sector	Private sector activities not regulated in Nairobi, Problems with one terms, conditions, and specification of contract signed,	Regulation and by-law exist, but weak enforcement in Dar es Salaam	Environmental Policy and by-laws exist and are enforced in Kigali,
Emerging PSI	High level of informality of open competition involving 60 private sector agents including few formal companies; only one contract as of 2002,	Five companies with contract in 1996, 65 private sector agents (CBO, NGO) as of 1999,	Four companies involved in franchised contracts, contract signed with households,
Level of service	House-to-house in middle/high income areas,	House-to-house in middle/high income areas, using trucks and push-cart,	House-to-house in middle/high income areas,
Private sector capacity	Mostly use conventional open trucks/ pickups; hire trucks; No standard solid waste vehicles	Mostly use conventional open trucks /pickups; hire trucks; No standard solid waste vehicles	Mostly use conventional open trucks /pickups,
Coverage and service quality	Low income areas not served, poor service quality; Unregistered CBO/NGOs disposed of waste indiscriminately,	Most households are satisfied with service by private organizations,	Households survey showed high service quality,
Cost recovery	$11-14 for door-to-door service by large private organization and $0.7-1.3 by small ones; Pay tipping fee at disposal site ($1.3 per truck),	Low level of charges; Insufficient funds; Receive payment from municipality for street sweeping,	Municipal authority don not play any role in user fee/tariff setting,
Households' involvement	Weak involvement of households	Lack of involvement of households and LA units at Ward Level,	Households are involved in contract agreement,

3.5 Theoretical Issues on Explanatory Factors of Private Sector Performance

In the previous section, some of the studies on cost efficiency and service quality improvements were reviewed. This section reviews the factors explaining private sector performance in both developed and developing countries, to identify explanatory factors to be explored in this study.

3.5.1 Factors Explaining Private Sector Performance

There are many cases of dramatic reductions in costs as a result of the involvement of the private sector (Cointreau-Levine and Coad, 2000). Private sector companies cannot always be more efficient without effort to apply private sector management techniques. It is useful to understand the reasons why private sector can be more efficient than the public sector or why some private sector companies can be more efficient than other private sector companies, since some companies have competitive advantage over others. When the reasons for higher private sector efficiency are understood, it is possible to prepare arrangements for involving private companies such that higher potential for efficiency is achieved in practice (Cointreau-Levine and Coad, 2000).

Private sector involvement in urban solid waste collection services seeks to improve service efficiency and increased investment. However, if the inter-organisational arrangements and conditions are not well structured, private sector involvement may not achieve either of these objectives, regardless of the market forces (Cointreau-Levine and Coad, 2000). Before the objective of improving service efficiency can be achieved, private sector companies have to improve equipment holdings, upgrade their knowledge and skills to rationalize operations and maintenance, and to be accountable, whereas the public sector has to ensure proper performance monitoring.

The econometric studies in solid waste collection in the developed countries identified some explanatory factors for private sector cost effi-

ciency. In the USA, Callan and Thomas (2001) found that the organizational form (private, public) does not influence costs in solid waste collection. Bel and Costas (2006) study of Spanish cities and towns found that
organizational form does not influence costs overall, but market concentration (the number of companies in the waste sector) creates problems for
competition. Dijkgraaf and Gradus (2003) show no difference between
public and private solid waste collection costs under competitive contracting among cities in The Netherland. Dijkgraaf and Gradus (2007) identified that cost savings from PSI erode over time. Only cities that recently
started PSI show cost savings, but cost savings from PSI appear to erode
over time, since there were no cost differences between cities that had
started PSI earlier and that of public provision. Ohlsson's (2003) study of
Swedish cities found private provision more costly than public provision
because of higher input and capital costs for private firms. A study by
Szymanski (1996) on 365 English municipalities found that although savings eroded over time, private provision costs were lower than public production.

The private sector efficiency and reductions in private sector prices are
a result of certain internal and external factors. Box 3.4 gives some examples of some factors explaining private sector efficiency (Donahue, 1989;
Bartone, 1991; Cointreau-Levine and Coad, 2000; Batley and Larbi, 2004).
Explanatory factors of private sector efficiency identified include competition (including managed competition), management flexibility, operational
capability (rationalized operations), accountability for results, clear performance measures and no political interference. Managed competition
involves direct competition between municipal solid waste departments
(public) and private sector service companies for a publicly competitive
tendered service contract.

The debate on private sector efficiency gains over public sector is inconclusive. The results from researches on private sector performance over
public sector delivery showed that efficiency gains are mixed due to some
explanatory factors (Donahue, 1989).

Box 3.4 Factors Explaining Private Sector Efficiency

Internal Factors (Private Sector Capacity)

- *Operational capability and productivity.* Comparatively high costs for the private sector in Quebec Province were due to older and larger equipment, larger crew sizes, and lower vehicle productivity (Cointreau-Levine and Coad, 2000). Private sector service costs in UK and USA studies have been at least 25% lower due to rationalized operations (Cointreau-Levine and Coad, 2000). Private sector service costs in Latin American Studies have been about 50% lower due to higher labour and vehicle productivity (Bartone, 1991). Private sector service costs in a Malaysian study have been at least 20% lower due to greater efficiency or productivity (Cointreau-Levine and Coad, 2000).
- *Scale of Operation.* Size of the companies and economics of scale influence their efficiencies (Donahue, 1989).
- *Responsiveness to users.* A private sector service provider is accountable to its customers by providing quality service, and is obliged to react if the customers are not satisfied (Cointreau-Levine and Coad, 2000).
- *Management flexibility.* The private sector is less restricted by bureaucracy in obtaining spare parts for repairs, and so that it can lease equipment when it is needed and subcontracts to meet peaks in demand. Private sector management has more flexibility to hire qualified staff, to pay staff according to their performance, to terminate the employment of unsatisfactory workers, and to adjust working hours according to service demand (Cointreau-Levine and Coad, 2000).

External Factors (Contracts and Regulation)

- *Competition and economies of scale.* Competition may not necessarily lead to cost savings. Dijkgraaf and Gradus (2003) argue that competition showed no difference in cost. Some studies suggest economies of scale influence service efficiency (Donahue, 1989; Boyne, 1998).
- *Market regulation.* The regulation of prices, and market entry and exit has influence on the private sector performance.
- *Regulatory capacity.* Weak regulatory capacity in developing countries for public services was identified by Batley and Larbi (2004). The private sector generally enjoys more freedom from political interference, so that it can optimize the size of the work force and the ratio of professional to operational staff, and to concentrate its resources on the service for which they are intended (Cointreau-Levine and Coad, 2000). If performance measure and conditions are specified in the contract or agreement, and the private sector operator is supervised effectively, good standards of operation can be achieved (Cointreau-Levine and Coad, 2000).

Source: compiled by Author with references indicated

The argument of the proponents is that the main reason for the efficiency gain is ownership and competition whereas the critics argue that there are more other factors which explain the efficiency gain. The factors identified from review by Donahue (1989) included organisational forms (public or private), economics of scale and competition. These factors determine the efficiency of the private sector involved in solid waste collection. Bel and Warner (2008) argue that the inconsistent efficiency gain (cost saving) by the private sector is due to inadequate theoretical frameworks based primarily on competition and ownership. Bel and Warner (2008) identified factors influencing efficiency as nature of service market structure (market concentration), service characteristics (types of services rendered), geographic location and regulation.

The results reviewed so far on service quality and efficiency gains by the private sector delivery over that of the public sector delivery are based on the theoretical frameworks, primarily on competition and ownership or organizational form (Bel and Warner, 2008; Carter et al., 1992) and internal management. The findings from the studies on private sector efficiency in the United States, Canada, and Great Britain and some developing countries showed that the improved efficiency was attributed to internal management and operational flexibility (crew size, labour conditions, vehicle routing and scheduling, technology choice, managerial capabilities), level of competition for the market, scale of operation, accountability for results, clear performance measures, no political interference, market concentration and regulation. The optimization of vehicle and labour productivities has priority in developed countries, whereas in developing countries it may not be a priority and therefore the desired quality of service and efficiency may not be achieved.

These factors explaining efficiency in the developed countries are identified and used as lens to explore the PSI situation in Ghana. This study seeks to explore the role of these explanatory factors of private sector efficiency and service quality in developing countries. The difference in efficiency gain due to organisational form (private sector over that of the public sector) is not the focus of this study, but the focus is on the differences within the private sector. Within the private sector at the firm level, there are differences in efficiency and service quality. In this thesis, the focus is on the firm level efficiency and service quality, and identifying the factors that explain the private sector performance in urban solid waste collection. The factors to explore include private sector capacity (scale of operation, skills/capabilities) and the private sector and market regulation (competition, accountability for results, political interference) and their implications for sustainable urban solid waste collection. The question this thesis seeks to address is: how can we improve the private sector performance to ensure sustainable urban solid waste collection in developing countries? In other words, how can private sector perform better in service provision in developing countries? The issue of efficiency and service quality is addressed by focusing on factors such as the skills/capabilities of the companies, scale of operation, and firm regulation and competition.

3.5.2 Regulation of Service Provision

Regulation is simply rules and laws that impose obligations or prohibit specific activities. Regulation controls and incentives have the tendency to influence private sector performance. The gaps between formal rules (regulation) and actual practice depending on institutional capacity and political context at the local level do influence private sector performance. Private sector involvement in solid waste collection calls for a greater and more efficient role of sector specific regulation. Chang (1997) defined regulation as government activity that is intended to affect directly the behaviours of private sector agents in order to align them with the 'public interest'. Regulation is also defined as "sustained and focused control, exercised by a public agency over activities of private sector which are valued by a community" (van Dijk, 2006).

The primary rationale for government regulation is to remedy market failures such as monopoly power, externalities and information asymmetries or deficiencies. In the industry where customers rarely have a choice regarding their supplier, the incentives for productive efficiency may be weaker. This can then be translated into higher prices for customers. One of the roles of regulation is to ensure that the prices charged to customers do not reflect the higher costs caused by firm inefficiency. In turn, this provides regulated firms with an incentive to achieve productive efficiency. Comparing the productive efficiency of different firms generally requires more than a simple comparison of unit costs, due to the differences in the operating environment faced by different firms.

The market failures are regulated through two forms of regulation of public services: regulation of the market conduct and market regulation (Ballance and Taylor, 2005). The regulation of the market conduct by independent regulator (formal regulation) concerns monopoly regulation. The regulation of market conduct constrains the behaviour of market suppliers in areas such as quality, pricing, and access to information. The monopoly regulation is strictly limited to service that cannot be unbundled. For monopoly regulation, there is 'competition for the market' where governments create competitive conditions through service contracts, leases or concessions, and firms compete not for individual consumers in the market but for the right to supply the entire market. The role of the government is then to regulate monopolies by means of some sort of incentive mechanisms and sanctions. The agents are given incentives to enhance the achievement of government objectives of efficiency and effectiveness.

Perhaps the most important aspect of market conduct regulation is the regulation of the prices charged by monopolists. The two major forms of price regulation that have been employed internationally are price cap regulation and rate of return regulation. For price cap regulation prices are controlled by a reference efficient firm's cost and adjusted by reference to increases in general price index. On the other hand, the rate of return regulation permits price increases to a level that allows a predetermined maximum profit margin on the investment.

There are four reasons for price cap regulation (Ballance and Taylor, 2005). First, price cap regulation provides strong incentives for productive efficiency. This is because the regulated firm can increase profits during the period of the price cap by reducing costs. At subsequent reviews of the price cap, these efficiencies can be passed on to consumers in the form of lower prices. Second, price cap regulation has been thought of as providing fewer incentives for overinvestment and inefficiencies. Third, price cap regulation provides the regulated firm with greater freedom over the structure of its charges. Finally, price cap regulation was considered simpler to implement than rate-of-return regulation. In practice, however, price cap regulation has proven to be more complex than originally envisaged since regulators have considered the underlying costs of efficient operation rather than using cost function developed over time.

The second form of regulation is the market regulation or regulation of the industry structure. The regulation of industry structure seeks to promote competition through regulation by contracts and legal restrictions and rules regarding both market entry and the shape of corporate entities operating in the market. For services that can be unbundled, regulation of industry structure is generally preferable to regulation of market conduct, as competition will maximise the choices available to customers and provide greater freedom for companies to innovate and offer new products. Regulation of industry structure is likely to be cheaper to implement, as the information requirements of structural regulation should be much less demanding. In practice, however, the two forms of regulation generally operate alongside each other.

Competition and Economies of Scale Arguments

Differences in costs under public and private production have been attributed primarily to competition. Regulation to allow competition is seen as central to ensure quality and efficiency gains, even with regulation of monopoly. Theoretically, competition encouraged private managers to keep costs down. Economic theory suggests that competition in the market or for the market is to promote private sector efficiency, innovation, and consumer responsiveness. Competition implies rivalry between alternative suppliers for a share of a market. The economic theory and public choice suggest that there is a positive relationship between competition and service performance. This proposition has been challenged in literature since transactions through markets are likely to work when the number of sup-

pliers is high and information cost and asset specificity are low. The measures of competition include the number of suppliers in a local market, the
distribution of market shares, rivalry between public and private suppliers,
and service managers' perceptions of the intensity of competition (Boyne,
2003).

Competition is a key feature underlying theoretical claims for costs
savings, but turns out to be problematic even in solid waste collection (Bel
and Warner, 2008). Where there are larger numbers of bidders, there were
higher cost savings (Gómez-Lobo and Szymanski, 2001). Szymanski and
Wilkins (1993) found similar results where there was 20% savings in the
first year, but these savings disappeared in 2 years, suggesting underbid-
ding by companies. Domberger et al. (1986) found that under competitive
contracting in Great Britain solid waste collection there was no difference
in public and private costs; but in places where there was no competitive
contracting, public costs were higher. These studies show that competitive
bidding do not always lead to cost saving. Other factors may explain cost
efficiency and better service quality of the private sector.

Bel and Warner (2008) argue that there is over reliance on the importance of competition in waste collection. However, empirical results suggest that competition *for* the market is not sufficient to ensure cost savings
sustained over time. But more recent evidence suggest that changes in
market structure of the solid waste management sector, where there was
significant consolidation and incumbency has led to erosion in cost savings
over time (Bel and Costas, 2006; Dijkgraaf and Gradus, 2007, 2008).
Competition for the market also eroded due to incumbency – contracts are
typically renewed as other providers exit the market. The potential competition-for-the-market is for the initial contract, but not for renewal. Bel and
Warner (2008) indicated that waste collection is characterized by weak
competition or collusion, because of the concentration in the market. For
waste collection, weak competition between firms erodes potential cost
savings. Their analysis shows that competition for the market is expected
but not typically found. The public versus private debate places too much
emphasis on ownership and competition, but Bel and Warner (2008) argue

that the primary attention should be given to economies of scale, market structure, regulations of incentives, and the contract specification.

Dijkgraaf and Gradus (2008) argue that regulatory structure matters for solid waste collection. These results suggest the importance of regulatory environment from an industrial organization approach. In most countries, there is a strong association between private provision and competition for the market through competitive tendering, and public provision without competition for the market. Typically, public provision is outside a competitive framework. The benefits of competitive contracting (increased efficiency) would come primarily with competition-for-the-market. Even though, monopoly provision would continue to be necessary due to economies of scale (Bel and Warner, 2008). Boyne (1998) argues that contracting may lead to higher efficiency, but this could be the results of economies of scale rather than competition. The theory of economies of scale suggests that efficiency is influenced by level of service output whereas public choice theory attributes efficiency gain to competition (Boyne, 1998).

Public Service Regulatory theory

Regulation of private sector involvement has an element of regulation 'by contract' where the regulatory rules are enforceable in a contract (Ballance and Taylor, 2005; Franceys and Gerlash, 2008). Regulatory theory originates from the works of Stigler 'capture' and Peltzman's general theory of regulation and the influence of 'public interest' (Hantke-Domas, 2003). There are two theoretical cases of regulation: the first is "agency capture" regulation – where regulation is a rent-seeking exercise dominated by special interests groups of those who regulate or regulated businesses, and the second strand is "public interest" regulation – where regulation is a tool to weigh business and customer interests to reach equitable results in the face of market failure (Teske, 2004). The 'public interest' regulation seeks to protect the public interest. Regulatory capture is an aspect of regulatory failure (Franceys and Gerlash, 2008). The regulatory 'capture' revolves around the degree to which regulatory decisions making equate the interest of the public and the regulated business, and the extent to which regulation of prices and service levels are specified in the contract or left to some form of extra-contractual resolution.

The current regulation debate is between those who call for more regulation of the private sector to protect 'public interest' and those who call for less regulation because regulation is most often 'captured by interest groups' and does not benefit the public. The key issue in this debate is whether regulation successfully serves the public interest or whether regulation benefits interest groups and consequently distorts the market and undermines efficiency. It is argued that public interest protection is the basis for the different forms of regulation – regulation of the market conduct and market regulation. However, others argue that regulation does not protect the public interest but only the interest of groups who capture the regulation (Hantke-Domas, 2003). Etzioni (2003) argue that regulatory capture is widespread and therefore instead of debating whether more or less regulation is needed, what is needed to be done is how to make regulation stronger and more capture-proof. Etzioni (2003) argues that those special interest groups who capture the regulation do so in a number of ways including: (1) fixing of higher prices, (2) they shape regulation by drafting rules, (3) dilute existing regulation with time, (4) weakening enforcement of regulation, and (5) repeal of existing regulation or altering of rules.

The regulation seeks to address market failures and gives a framework for analysing regulation of PSI. Governments regulate market structure in various ways in order to overcome market failures. Market structure refers to the number of firms in a sector and the nature of their interactions. The ways of regulating market structure include removing barriers to entry, restrictions on market concentration, and restrictions on vertical integration. Governments may also regulate market conduct, which includes controlling operators' pricing and practices or providing incentives for appropriate conduct. Regulation of market conduct is traditionally viewed as a poor substitute for competition (Teske, 2004). As a result, regulators often encourage competition whenever practicable. The advantages of competition over regulated conduct include limited opportunities for political rent seeking and collusion, fewer information asymmetries, and better incentives to serve customer interests.

There are two basic schools of thought on regulatory theory; namely, positive theories of regulation and normative theories of regulation (Jamison and Berg, 2008). Positive theories of regulation examine why regulation occurs. These theories of regulation include theories of market power, interest group theories that describe stakeholders' interests in regulation, and theories of government opportunism that describe why restrictions on government discretion may be necessary for the sector to provide efficient services for customers. In general, the conclusions of these theories are that regulation occurs for three reasons. First, regulation occurs because government is interested in overcoming information asymmetries with the operator and in aligning the operator's interest with the government's interest. Second, the customers desire protection from market power when competition is non-existent or ineffective. Customers would like to see regulation of user charges and quality of service. Finally, regulated businesses desire protection from rivals and also protection from government opportunism.

Normative theories of regulation generally conclude that regulators should encourage competition where feasible; minimize the costs of information asymmetries by obtaining information and providing operators with incentives to improve their performance; provide for price structures that improve economic efficiency; and establish legitimacy of regulatory processes that provide for regulation under the law and independence (Jamison and Berg, 2008).

This study seeks to explore regulatory mechanisms, the influence of regulation on performance, and the legitimacy of regulatory process in solid waste collection. The legitimacy of a regulatory regime in this sense refers to its acceptance by stakeholders, such as customers and regulated businesses, as a source of authority that exercises its powers in a fashion they consider appropriate (Balance and Taylor, 2005). Baldwin and Cave (1999) set out five indicators that focus on the issues of regulatory processes to test the legitimacy and performance of a regulatory regime. These indicators are: (1) the political or legislative mandate of those making regulatory decisions, (2) the perceived 'fairness' of the decision-making processes and involvement of stakeholders, (3) the accountability of regulatory decision makers, (4) the regulatory capacity and expertise of decision makers and, (5) the efficiency and overall cost of the regulatory regime. Balance and Taylor (2005) argue that the perceived legitimacy of a regulatory regime in the eyes of those stakeholders who are subject to its will, have an important influence on the stability and longevity of that regime. This perceived legitimacy is likely to be influenced by a number of factors, including not only the outcomes of the regulatory regime, such as the level of prices and services received, but also on the five indicators of legitimacy of regulatory process.

Regulatory Impact

The impact of regulation on service performance may be positive depending on the regulatory instruments and the autonomy of the regulator, and regulatory arrangements. The evidence of the impact of regulatory arrangements for service performance is clearly sparse, weak and incomplete (Boyne, 2003), and the gap yet to be investigated is the impact of the nature and number of regulatory instruments. There is a range of regulatory instruments that include audit, inspection, financial control, performance indicators, plans, and annual reports (Boyne, 2003).

In contracting out public services, the principal-agency theory suggests that it is imperative to separate the identity of the principal (public agency, local government department) from the agent (private sector), and thus monitoring and reporting become outcome whether the services are delivered by the public bureau or through the market. The separation of regulatory tasks from service provision contributes to the positive impact of regulation on service delivery. This is done through separation of the purchaser (public sector agent) and the provider of the service. Regulation mechanism, with separation of purchaser and provider, affects the extent of efficiency and effectiveness in service delivery.

The outcomes from government regulation of a market are not always those that are intended due to principal-agent problems, regulatory capture and problem of legitimacy of entire regulatory regime. The principal-agent problems can occur where regulatory agencies have different objectives or incentives to pursue that are different from the elected politicians (principal). Related to principal-agent problems is the problem of regulatory capture, where the regulatory agency does equate the interests of society with those of the firm they are regulating (Ballance and Taylor, 2005). Regulators may be impeded in their work as a result of information asymmetries whereby the regulated company has greater knowledge of costs than the regulator, which it can use to its advantage and thus distort regulatory decision making. The legitimacy of an entire regulatory regime may come into question if the processes they employ are not transparent, if decision makers are not accountable and if those affected by regulatory decisions do not have sufficient opportunity to have an input into the decision-making process.

The extent of regulation is affected by cost of monitoring contract. The transaction costs are the cost of reaching and enforcing agreements (Field, 1997). Transaction costs relate to all expenses regarding a contract: the costs of searching for contractual partners, acquiring information and negotiating; monitoring and enforcing contracts. The transaction costs may be high or low depending on human factors (rationality and opportunism), uncertainty, and information asymmetry. The concept of transaction costs has contributed a great deal to the problem of whether to contract out or establish a bureau (public sector department) to deliver services (Awortwi, 2003).

3.5.3   Internal Factors of Private Sector Performance – Capacity of companies

Scale of operation

Company size is conceptualized as a scale of operations. The measures of organization size include the number of solid waste collection vehicles, the number of staff, and the number of service users. There is little empirical evidence that there is positive linear relationship between performance and size. There is size effect on performance of firms, but in most researches size effect is not the focus of the studies and as such it is used as control variable. Company size has relationships with performance. However, empirical studies testing the relationship between size and performance had yielded mixed results due to the differences in how performance and size were measured (Nachum, 1999). There is little empirical evidence that the relationship between performance and size is linear. Boyne (2003) argued that there is non-linear relationship between organisation size and performance, since medium size organisation may perform better than the small or large organisations.

Private organizations providing public services may be small or large in size and internal structure may be formalized (e.g. reliance on rules) and centralized (power and span of control). The benefits of large organisation taking advantage of the economics of scale and scope could be compared with the benefits of small organisation. Public choice theorists suggest that the fragmentation of large organisation result in responsiveness and allocative efficiency (Boyne, 1998).

Stevens (1978) found that better technology (and larger trucks) among private solid waste collection providers in large cities enabled them to use smaller crews which contributed to efficiency. These differences increased in magnitude with city size. According to Stevens (1978) scale economies seem to be exhausted at a relatively low population level (20,000–50,000). Dubin and Navarro (1988) found economies of density in solid waste collection but not scale economies. Empirical evidence testing the relationship between size and performance has yielded mixed results (Nachum, 1996).

Skills and Capability for Operations Management

Organizational capability is defined by Grant (1991) as a firm's ability to perform repeatedly a productive task which relates either directly or indirectly to a firm's capacity for creating value through effecting the transformation of inputs into outputs. The internal capacity (human resources, machinery capacity) of an organization affects or shapes the organization's performance (Lusthaus et al., 2002). The simple theory of public service improvement is that more resources lead to better results. But the resources must be effectively managed in order to deliver the maximum potential benefits. According to public choice theorists, budget maximization leads to allocative efficiency (meeting demand) but not productive efficiency. The quality of service can be expected to deteriorate as the sizes of the public budget expand. However, in the private sector, extra resources couple with prudent operations management is likely to lead to improved performance in order to maximize profit. Financial resources allow real resources to be purchased and this in turn influences service performance. The proposition is that the relationship between resources (financial and real resources) and service improvement is positive (Boyne, 2003).

The managerial task is very demanding and essential for success. The need to produce creative work requires special managerial efforts and capabilities. The impact of managers on their firms' performance is likely to vary depending upon such factors as the knowledge, intelligence, and experience of the persons holding the managerial position. The management literature suggests several measures to capture the relative competence of managers and among them are: the level of qualifications, number of years of experience, lengths of managerial experience (Nachum, 1996).

The quality of employees depends on the individual knowledge (creation of manuals, expert systems, team work, business code and so on), which is transformed into organizational knowledge. The quality of the employees could be measured by the salary level. As an operational measure for the quality of employees average salaries are used as proxy (Nachum, 1996). Using higher salaries to attract the best talents has become a common strategy in many industries.

Management (operational and strategic processes) in private organisations has relationship with performance of the companies (Hansen and Wernerfelt, 1989; Boyne, 2003). The management variables include both strategic variables (strategy processes and content, leadership styles, and human resources management) and operational/tactical variables (capacity planning, operations supervision, operations improvements, and service design and maintenance management).

3.6   Enabling Environment for Improving Service Performance

Having reviewed the conceptual and theoretical issues of private sector involvement in service provision and the factors which influence private sector performance, this section seeks to identify the elements of enabling environment (institutional capacity, policies, legal, regulation) and private sector strategies and innovations that help to increase private sector performance. The review of private sector performance, capacity and regulation suggest that some basic fundamentals for service provision need to be right in order to achieve good performance of the private sector, better service provision, and the ultimate goal of sustainable solid waste management. There is the need to provide enabling environment geared towards improving private sector performance (productivity and service quality). The private sector also needs to take competitive advantage of the private sector management strategies and innovations. It is necessary to get the fundamentals right and go beyond the basics (waste collection) to improve service provision and to work towards sustainable solid waste management.

3.6.1   Theoretical Case of Enabling Environment for Service Provision

Public sector capacity for Public Service Provision

UNDP (2008) defined enabling environment as the elements at the national or local level that can facilitate or constrain the development of capacity. These include policies, rules and norms, values governing mandates, priorities, modes of operation, and culture. Such elements exist within and across sectors of economy. They create incentives or constraints that determine the 'rules of the game' for interaction among sectors and can affect the performance of one or more organizations within a sector. The enabling environment (institutional capacity and arrangements) for regulating solid waste collection concerns the presence and use of polices, strategies, legal, regulatory rules, political will and commitment to further advance the development processes in solid waste management. This section presents the context of enabling environment necessary for better service provision and improved service quality.

The hypotheses for this study are derived from the review of internal and external factors of private sector performance and the necessary capacity conditions necessary for improving private sector performance. The findings of this study will be compared with theory to draw theoretical implications and conclusions. To examine the theoretical implications of this study, the theoretical themes presented in this chapter will be used as lens to view and examine the findings of this study. The three theoretical themes of this study to help understand the local context and the scheme of study are: enabling institutional capacity and regulatory rules and contract, private sector performance and capacity, and user involvement in private provision of public services.

The public sector agents have an important role to play in order to ensure equitable access to services. They must provide the policy, legal and regulatory frameworks which can maintain the necessary levels of service quality and accessibility. This could be done by regulations, changes in pricing and subsidies, but also through private sector involvement and participatory approach (Gidman et al., 1999). Service delivery through orientation on the demand side needs institutions that have the capacity to identify and manage that demand and then be more accountable to users and responsive to their needs in delivering the services needed.

The fragmentation of service areas to promote multiple agents involvement could also provide competition and improved service provision. In order not to create monopoly situation, solid waste collection area is divided into segments with separate contract for each and with at least one segment being retained for public management. Such a procedure ensures that the public organisation always retains the ability to provide services directly and that service delivery standards do not deteriorate or unfair pricing practices develop. The local government capacity – the ability to undertake and promote collective actions efficiently – must be increased. Matching the local government's role to its capability is necessary for service provision (World Bank, 1994).

Regulation and Provision of Incentive Mechanisms

The causes of past poor performance, and the source of improved performance, lie in the incentives facing providers (World Bank, 1994). To ensure efficient, responsive delivery of services, regulation (through legislation and incentives) need to be introduced through the application of three instruments – commercial management, regulatory mechanisms and competition, and stakeholder involvement (Gidman et al., 1999). Well-designed regulatory systems can help societies influence market outcomes for public ends and protect consumers, workers, and the environment. Regulation can foster competition and innovation while constraining the abuse of monopoly power (World Bank, 1997). Zoning of solid waste collection area with separate contract for each zone and at least one zone retained by public organization with adequate capacity provides the ability of public organization to ensure fair pricing practices and that service delivery standards do not deteriorate. The need for private enterprises to make profit may lead to user fees which are too high for low income groups, unless user fees are regulated and cross subsidisation provided.

The private sector is not politically accountable for public interest but it is regulated by the public sector to safeguard public interest. For private sector to involve customers (customer orientation and households' involvement) and to be responsive to users and to be accountable for results, it is regulated using regulatory instruments (incentives and legislative powers of control) (World Bank, 1994; Gidman et al., 1999). The increasing involvement of private sector without commensurate households' involvement (user charging, education) in service provision does not ensure customer-oriented service and better service quality.

Service delivery through orientation on the demand side (customer-oriented service) needs service providers that have the capacity to stimulate and manage demand and then be more accountable for results to users and responsive to their needs in delivering the services (Gidman, et al., 1999). Clients can be involved in specifying their need (choice) and in the effective monitoring of the service providers as they are at the point of service delivery (World Bank, 2004:9). Figure 3.2 shows the typology of regulation (weak or strong) and service quality (low or high).

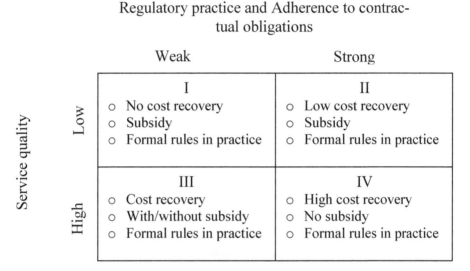

Regulatory practice and Adherence to contractual obligations

	Weak	Strong
Low	**I** ○ No cost recovery ○ Subsidy ○ Formal rules in practice	**II** ○ Low cost recovery ○ Subsidy ○ Formal rules in practice
High	**III** ○ Cost recovery ○ With/without subsidy ○ Formal rules in practice	**IV** ○ High cost recovery ○ No subsidy ○ Formal rules in practice

Service quality

Figure 3.2 Typology of capacity, customer orientation-service quality relationship

When the whole service provision is privately run, the public agency retains a responsibility to safeguard public interest. This is normally achieved through such regulatory instruments and customer watch dogs set up to regularly scrutinise the private sector performance and adherence to fair pricing and quality. Ideally such regulatory arrangements will have legislative powers of control built into their constitution (Gidman et al., 1999). The new role of the local government agencies should be flexible, facilitating and focused towards active participation of all actors involved and provide social, financial and technical support to the urban poor (Gidman et al.,

1999). If the local government does not play that role, the private sector will concentrate only on high or middle income areas leaving the poor out to rely on more expensive private suppliers or resort to indiscriminate dumping of solid waste. The local government can play a fundamental role (creating an enabling framework of laws, economic and political conditions) in helping private sector agents to play their roles more effectively and as a result increase the access to services for the urban poor.

Regulation of prices by the local government where service charges and collection fees are sufficient to cover full cost of service, where investment in new collection vehicles is recovered, result in better service quality of the companies. This is because the setting of economic service charges and allowing cross subsidisation for the urban poor provide incentives for the private sector to recover investment and provide better service. Cost recovery is high in communities where there is high level of households' participation in the service provision. This is because households' involvement and orientation motivate the households and increase their willingness-to-pay for services.

3.6.2 Private Sector Capacity, Responsiveness to Users and Accountability for Results

Besides enabling environment, private sector companies have a role to play to be competitive and to stay in business amidst regulated prices and service quality. They must have the adequate capacity for service provision, and the fact that the private sector has to make profit helps to ensure that resources are used more efficiently and also stimulates innovations and competition among the different private partners (Gidman et al., 1999). In theory the performance of a firm is linked to the availability of physical and human resources.

The private sector can provide better service quality when service providers show responsiveness to users and accountability for result (Gidman et al., 1999). The establishment of public private partnerships can be used to make organisations more accountable. Public private partnerships are also a means to create competitiveness and allow for more effective articulation of demand. In order to balance cost recovery with equity two conditions are essential. First, people will not pay for services they find inadequate, so there must be a strong link between user charges and quality perceived by users. Second, cost recovery is easier where there is high level of participation by the users in the management processes. The user participation promotes cost recovery and provision of better service quality.

3.7   Research Hypotheses and Theoretical Framework

3.7.1   Research Gaps and Issues

The review of literature on market mechanisms for public service and the enabling environment needed for private sector performance raises the following research gaps and issues which are relevant for developing countries.

o   *Private sector performance.* Results on public sector efficiency over private sector efficiency in developed countries are mixed. The variations in private sector efficiency in developing countries could be explored to find out the factors explaining private sector performance in developing country context.

o   *Scale of operation and economies of scale.* Scale of operation (and economies of scale) was seen as an issue in studies that focused on this variable. Efficient use of input resources depends on scale of operation and volume of waste to be collected in a given zone, but the minimum optimum scale of operation is least explored in developing countries.

o   *Internal management and technical capacity.* Most of the studies in developing countries did not focus on internal management of the companies. Those companies in developed countries that had lower costs may rely on optimizing operations and internal cost savings.   The internal factors which influence private sector performance in solid waste collection are least explored in studies in developing countries.

o   *Regulation*. Regulation of the private sector companies involved in solid waste collection is least explored in developing countries. Theoretical framework for analyzing regulation provides the basis for this study. The role of contracts, competitive tendering, regulatory instruments and regulatory capacity are explored. What happens if there is competition for initial contract but no competition for subsequent contract renewals?

Within the private sector at the firm level, there are differences in efficiency and service quality. In this thesis, the focus is on the firm level efficiency and service quality, and identifying the factors that explain the private sector performance in urban solid waste collection. The factors explored include private sector capacity (scale of operation, skills/capabilities), regulation (contracts, competition, accountability for results, political interference) and their implications for sustainable urban solid waste collection. The next section outlines the questions to be addressed in this thesis.

3.7.2   Research Hypotheses

Based on the literature reviewed and the Ghanaian context the main hypothesis that will be tested in this research is:

Private sector companies, with adequate capacity for service provision, are likely to achieve high performance when strong regulation practice exists and when formal rules and contractual obligations are strictly adhered to.

By contrast, the alternative hypothesis would be:

Private sector companies, with inadequate capacity for service provision, are likely to achieve low performance when weak regulation practice exists and when formal rules and contractual obligations are not strictly adhered to.

In order to test this main hypothesis, the following sub-hypotheses (key prepositions) are tested in this study. These propositions show how the various explanatory variables (internal and external factors) influence private sector performance. The specific hypotheses, based on overview theory on factors that drive performance, are as follows:

o   High productivity of a company leads to better service quality, and that high performance is achieved when both productivity and service quality are high.

o   The increasing involvement of the private sector coupled with households' involvement in user charging, service quality monitoring and waste minimization programme leads to customer-oriented service and better service quality.

o   Strong regulatory practices and adherence to formal rules and contractual obligations provide incentives for cost recovery and better service quality.

o   Scale of operation (size) of private companies and the capability for solid waste collection influence private sector performance (productivity and service quality).

The sub-hypotheses (key propositions) are discussed further in the next section under the theoretical framework.

3.7.3   Theoretical Framework

The theoretical framework proposed for this study to focus the research shows the relationship between performance of the service providers and the determinants of performance. This research seeks to assess the performance of solid waste service using two performance indicators: productivity, and quality of service. In the service sector, productivity is a measure of service efficiency whereas quality of service is a measure of service effectiveness. The cost efficiency of the business measures the financial sustainability of the service provision.

It is expected that the *skills and capabilities* for operations management (operations management, supervision and planning, and vehicles maintenance management), *scale of operation* (number of vehicles, market share of waste collected), and *contract and regulatory mechanisms* (access to information, user charges, quality of service, contract obligation) will cause differences in performance of the private service providers since each city is under different Local Government but implementing the same environmental sanitation policy. To investigate these differences, comparative analysis between companies in the same city and five different cities is made using both quantitative and qualitative methods.

Dimensions of performance

Performance assessment approaches focus on economy, efficiency and effectiveness (service quality) of service delivery (Boston, 2000; Kessey, 1995; Carter et al., 1992; Cointreau-Levine and Coad, 2000). There are inherent difficulties in measuring quality of many services if relevant benchmarks are not established (Boston, 2000). Osborne and Gaebler (1993) argued that, organisations that measure the output or service quality of their work are able to improve performance. The choice of dimensions and determinants of performance depends on the perspective of the researcher.

Productivity

Productivity is defined as the ratio of outcome (or output) to input resources used. For example the ratio of waste collected in tonnes to a vehicle, where output is the waste collected and input is the vehicle used. The choice of vehicle, capacity and physical conditions of vehicles affect productivity. There are relationships between vehicle productivities, operations management and the cost of services. The number of vehicles and capacity do not say much when comparing performance of companies unless productivity, companies' capacity and waste generated in the designated areas are compared. In solid waste service, the extent of vehicle utilization also affects the productivity of the companies.

Productivity is also defined as productive efficiency because it results in cost savings that directly contribute to profitability, competitiveness, and corporate survival. In the service sector, firms have strong incentive to achieve maximum productivity or productive efficiency to reduce cost and gain market share. Most productivity improvement efforts in for-profit organizations focus on ways to increase employee output (or labour productivity) and speed up production.

In residential refuse collection, for example, useful output measures are the number of tons collected, and the number of households serviced (Savas, 1978). These lead to the efficiency measures of cost per ton, cost per household per year, tons collected per man hour, and households serviced per man-hour. Much effort has been devoted by management scientists to improving the efficiency of refuse collection, ranging from changes in institutional arrangements for providing the service, to changes in vehicles, in vehicle routing, in manpower scheduling, and in siting of transfer stations or disposal points.

Quality of service

Service effectiveness is defined as the level of outcome accomplished, for example, the level of satisfaction that customers experience with services, the number of waste pickups made, the amount of revenue collected from customers, and the number of customers that are served. In the service sector, quality of service is used as a measure of service effectiveness. Effectiveness is another important measure of performance. It measures how well the need for the service is satisfied and the extent to which unintended adverse impacts are avoided. It is a measure of the adequacy of service relative to need, and incorporates the notion of service quality (Savas, 1978).

In this study, quality of service is used as a measure of perform-ance of solid waste service. The measures of quality of service used in this study are reliability of service, customer satisfaction, service coverage, re-sponsiveness to customers, and environmental cleanliness. These measures are the objectives of the service to be met in practice. The objectives which are the service outcome or quality standards are usually set by the principal partner of the service contract, for example the Waste Manage-ment Departments in the Ghanaian cities. The distinction between output and outcome is relevant because organizations often have more control over output than outcome, because outcome depends on the perception of the customers whereas output is easily measured. For example, a waste com-pany collects certain amount of waste in a day (output) which in turn makes the city clean (an outcome).

Measuring service quality is challenging because customers perceive the service based on their prior expectations and compare these with the perceptions of the actual service. Customers are delighted with the service when the perceptions exceed expectations, and therefore the understanding of customer needs and service objectives are important. The objectives (service quality standards) concern the extent to which outcomes are achieved, the extent to which predetermined standards are met, reliability of the service (i.e. dependability), timeliness of the service and good aes-thetics (Folz and Lyons, 1986). Service quality assessment requires that customers know all the service quality standards that should be met by the service providers. Customers are more concerned with the quality of ser-vice than efficiency; since efficiency gain may not necessarily translate into effectiveness.

Explaining the relationships between performance and its determinants

Scale of operation

There are differences in the manner in which company size is operationalized in the literature. Company size is conceptualized as scale of operations, but there are no universally accepted criteria for measuring company size. The most common measures of company size in the service industry are: the number of daily outputs, the number of employees, the level of sales, and the value of firm assets. Economies of scale give companies with scale of operations above the minimum efficient scale potential advantage to render services at lower average costs. The minimum efficient scale of Private Sector Company is important for efficient and effective service delivery.

It is postulated in this study that scale of operation has influence on productivity. An optimum scale of operation can lead to productivity and cost efficiencies. The question this study addressed is how does scale of operation varies with productivity of companies involved in solid waste collection? Could high productivity in solid waste collection be achieved through optimization of scale of operations and utilization of capacity of collection trucks?

Skills and Capabilities

Operations management capacity affects organization's performance of service operations. The measures of operations management capacity include the quality of employees, both technical and managerial, and the resources management ability (Lusthaus et al., 2002). Organizational capability is defined by Grant (1991) as a firm's ability to perform repeatedly a productive task which relates either directly or indirectly to a firm's capacity for creating value through effecting the transformation of inputs into outputs.

There is a relationship between capacity for operations management of services involving especially logistics and performance (Hansen and Wernerfelt, 1989; Boyne, 2003). In exploring the impact of operations management capacity on performance and identifying possible processes of influence, it is necessary to investigate the performance relationships using measures of operations management capacity. In this study, the measures of operations management capacity used are operations planning capacity, maintenance management capacity, span of supervision, and quality of employees. The operations planning capacity involves the ability of the operations manager to plan waste collection and establish rational vehicle routes, route plans and vehicle logs for the service operations. The maintenance management capability is the ability of the technical personnel to plan in-house maintenance, prepare maintenance schedule to prevent equipment breakdown and long downtime duration. The operations supervision involves the ability of technical personnel to monitor collection services and emergencies.

Regulation Rules and Contract

A comparative analysis of regulatory mechanisms requires a testing of hypothesis based on regulatory framework. The theoretical case that guides the hypothesis to be tested is focused on local government interest, public interest and private sector interest and these are combined to test relationships suggested by the agency capture and public interest theories of regulation. The theoretical framework for analysing regulation is based on market failure framework (regulation of monopoly power, asymmetric information, and externalities) and a contrasting theoretical case of "public interest or agency capture" perspective for explaining regulation.

An inter-organisational arrangement that has gained prominence in recent years is the external structure through which services are provided (Boyne, 2003). In this arrangement, the public agencies may be part of a partnership or may contract out parts of their services to other providers. The performance of this arrangement will depend on whether the organisations involved are able to mobilise more skills and resources than a single organisation delivering the service. The key dimensions of the inter-organisational arrangements include size of delineated zone to service, zonal monopoly, type of contract, contractual obligations. A contract relationship between two partners starts with the signing of a contract with clear specification and service characteristics which set out the minimum acceptable service performance, contract duration, and obligations of parties.

3.7.4 Framework for Analysing Sustainable Solid Waste Management

Solid waste management is one of the most important services for ensuring sustainable development and maintaining the quality of life in urban areas. Sustainable solid waste management involves a system that is appropriate to the local conditions from a social, financial, and environment perspectives, and capable to maintain itself over long time without reducing the resources it needs (van de Klundert and Anchutz, 2001). Sustainability as used in literature has a broader meaning and application in development studies, environment or business depending on the focus of what has to be sustained – project, system, process, activity, or resources. Sustainability as defined in the field of development studies is the capacity of a project or activity to continue to deliver its intended benefits over a long period of time (Kasim, 2006). Sustainability of a system is achieved when it is able to deliver an appropriate and equitable level of benefits in terms of service quality and affordability over a prolonged period of time without negatively affecting the environment (van de Klundert and Anchutz, 2001). This implies that the beneficiaries are satisfied with the service and the costs are covered through user fees or financial mechanisms.

Solid waste collection encompasses three important interrelated aspects (environment, financial and social) that hinge on sustainability. Environmental sustainability requires that solid waste collection and disposal which impose great burden on the environment and resources, be transformed into a closed-cycle system (closing the loop) by restoring various natural cycles, thus preventing the loss of raw materials, energy and nutrients. This means that waste reduction and minimization are integral part of solid waste collection through source separation and separate collection. Financial sustainability ensures that there is a more sustainable cost recovery approach than reliance on government subsidy financing. Full or partial cost recovery through user charges based on ability-to-pay reduces the financial burden on the government. Social sustainability of solid waste collection concerns providing services to all strata of society, regardless of income. The measures of the environmental, economic and social sustainability of solid waste collection are presented in Table 3.3.

The framework for analyzing sustainable solid waste collection and management (Figure 3.3) is derived from a combination of development cycle of Wilson et al., (2001) 'strategic planning guide for sustainable municipal solid waste management' and frameworks by Cointreau-Levine and Coad (2000), Lusthaus et al. (1999), Lusthaus et al. (2002), and Batley and Larbi (2004). Wilson et al., (2001) provide four stages of ideal development cycle for improving solid waste management: (1) institutional development and service performance; (2) improved customer satisfaction and public awareness; (3) improved cost recovery and financial performance; and (4) sustainable levels of investment.

Table 3.3 Measures of sustainability of solid waste collection

Sustainability	Indicators of sustainability
Environmental sustainability	Zero uncollected waste (100% collection) and households willingness to source separate waste, recycle and reuse waste,
Financial sustainability	Full cost recovery through user charges and pay-as-you-throw (PAYT) and little or no government subsidy,
Social sustainability	100% service coverage (as a measure of equity and fairness to inhabitants)

This framework combines institutional (policy, legal) context with interaction between public sector and private companies as well as community participation. The institutional development and service performance stage is concerned with institutional frameworks (policy, legal, institutions) for solid waste management and service performance monitoring.

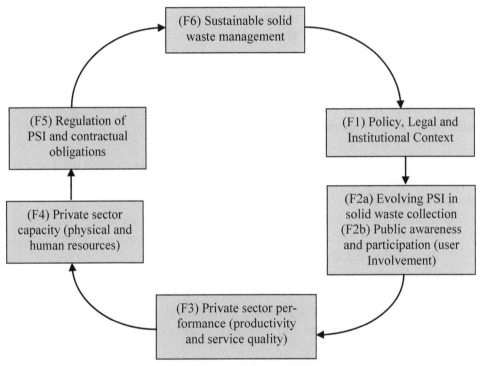

Figure 3.3 Conceptual Framework for analyzing Sustainable Solid Waste Collection

Sources: Framework derived from Wilson et al., (2001)

The framework for analyzing sustainable solid waste management has six components (Figure 3.3). The first component (F1) is the policy, legal and institutional frameworks for solid waste collection. Institutional arrangement for solid waste collection concerns the institutional structures

and measures as well as organisational procedures and the capacity of responsible institutions (Schubeler et al., 1996). The institutional arrangements seek to detail the policy, legal and institutions for service provision
(Wilson et al., 2001). These arrangements characterize and provide frameworks for the role of the private and public sectors stipulated in the policy
and legal frameworks.

The second component (F2 in Figure 3.3) is the evolving PSI and role
of public and private sectors in practice. This component which focuses on
the evolving involvement of the private sector will be explored in chapter
5. The focus of the 'evolving involvement of the private sector' and the
public-private interactions will be on how PSI started, why PSI happened,
how PSI organised in the past and now and what are the conditions for
success. These questions would be used to explore the evolving involvement of the private sector.

The following dimensions of how PSI was organised were used to explore the evolving involvement of the private sector and households over
two decades (1990-2010):

o *Zoning of city into service areas for more private sector participation* (size of collection zones, and zonal monopoly – comparison of
  population per collection zone across cities),
o *Evolving forms of public-private arrangements that exist in practice in the five cities from 1990 to 2010* (number of companies, nature of contract or licensing, companies' cost recovery arrangements – user charges and subsidies – and companies' market shares
  of waste collected),
o *Trend in service coverage and level of service* (proportion of house-
  to-house collection, proportion of communal collection, public collection, private collection, collection rate),
o *Households' involvement* in service provision and awareness of
  solid waste impact,
o *Implementation of national policies over two decades (1990-2010),*
  and implications of current practices for policy and regulation at local and national level.

The second component also concerns public awareness and participation. The analysis of the public awareness and participation component seeks to explore whether resident were involved in the service provision arrangements. The interactions or relationships among public-private-community are seen as contributory factor to the success of service provision and promotion of willingness-to-pay for service. The pace of development of waste management services depends on the level of public awareness and participation to achieve improvements (Wilson et al., 2001). The public support for any issue can be greatly increased if the public is fully informed of the reasons behind their action, and the intended benefits. The analysis of public awareness and participation in this study (in chapter 5) covers awareness of waste-related public health, public participation and support for service provision and cost recovery, willingness-to-pay for collection service, and willingness to support recycle and source-separation activities.

The third component (F3 in Figure 3.3) is the analysis of private sector performance. The analysis of private sector performance (in chapter 6) is intended to provide insight into the level of productivity and service quality. The level of performance is used in analytical analysis in chapters 7 and 8 to explore the factors which explain private sector performance.

The fourth component (F4) is the analysis of private sector capacity. The analysis of private sector capacity is intended to provide insight into factors driving or affecting performance so that performance improvement measures can be implemented and monitored. The analysis of private sector capacity in this study (in chapter 7) covers scale of operation and human resource capability.

The fifth component (F5) is the analysis of regulation and the contractual rights and obligations of public and private sectors. The analysis is presented in chapter 8 and covers the elements of regulation, contractual rights and obligations including the following:

- o Contract specification and scope of services
- o Competitive tendering and level of competition
- o Contract duration, renewals and termination
- o Regulation of access to information

o Regulation of unit prices, user charges, payments and cost recovery,
o Regulation of service quality (performance monitoring and sanction of noncompliance)
o Regulatory capacity and autonomy,

Finally, the sixth component of the framework (F6) seeks to examine the components from 1 to 5 to identify the drivers for sustainability solid waste management. The identification of drivers for sustainability is intended to provide feedback for changes in policy and legal frameworks.

3.7.5 Analytical Framework

This section further outlines the indicators for assessing the performance and capacity of organizations within the context of their external environment (regulations) and the evolving processes of households' involvement. The analytical framework is shown graphically in Figure 3.4. The analytical framework provides the indicators for assessing the performance of service providers and its determinants. Because performance is relative to an organization's basic capacity, the analysis of capacity sets the stage for understanding organizational performance (Lusthaus et al., 1995).

The framework provides the basis for operationalising the performance and its determinants. The determinants of performance are the internal and external factors which influence private sector performance. To explain the differences in performances between private sector organisations, a framework which combines three concepts: capacity (physical and operations management capacity), regulation (contractual relations and regulatory mechanisms) and evolving practices of involving private sector and households in service provision is used to explain the variations in performance. The key indicators of performance and determinant of performance are presented in Figure 3.4. The figure shows detailed list of indicators and variables to be used in the assessment of performance and determinants of performance. The variables in the analytical framework for the study are categorized into two: (1) Dependent Variables: performance of companies in terms of productivity and service quality, and (2) Independent Variables: determinants of performance of service delivery. There are indicators for measuring each variable in the framework. The key variables of this framework are operationalised in the next chapter.

The operations management capability of private firms focuses on the management of operations using the available human and physical resources. The physical capacity (scale of operation) of the companies as a variable has a relationship with performance. Both company capability (human resources) and physical capacity are the internal factors that may influence performance. The external factors concern with the regulatory mechanisms and the processes of evolving involvement of households for efficient and effective urban solid waste service.

The regulatory variable seeks to inquire into the types of contracts, regulatory mechanisms, contractual obligations and relationships, and cost recovery in the public-private partnership. The variables of evolving practice of involving private sector and households in service provision seek to assess how the changing role of the local government and its processes have influenced solid waste service provision and private sector performance.

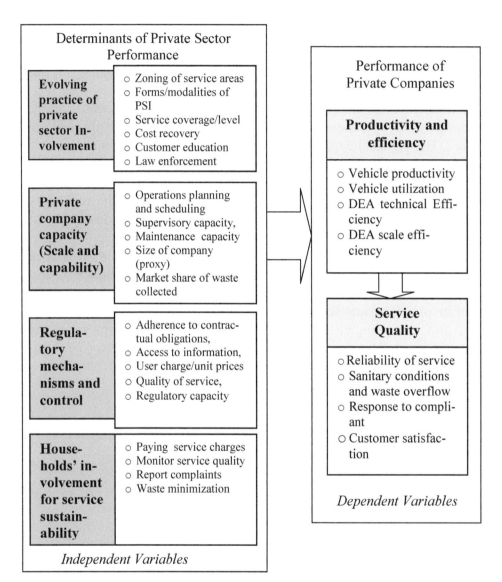

Figure 3.4 Analytical framework

3.8   Conclusions

In this chapter, five theoretical issues on private sector involvement in service provision were reviewed: (1) government and market failures, (2) theoretical debate on private sector involvement in service provision and performance, (3) theoretical debate on explanatory factors of private sector performance, (4) enabling factors and drivers for improving performance, and (5) framework for analyzing private sector involvement and sustainable service delivery. The review of government and market failures, as well as the reasons for private sector involvement in service delivery provides the theoretical context for the study. There is a shift in the role of the public sector, with the public sector taking a leading role as regulator of market to correct market failures. The private sector is now taking on an increasing role as provider of public services.

Based on the review in this chapter and the research problem in chapter 1, the theoretical and conceptual frameworks and hypotheses for the study were presented. The theoretical framework for analyzing the factors which influence private sector performance in service provision will be operationalised in chapter 4 and variables to be measured defined. Some empirical studies in developed countries were reviewed to identify the factors which influence private sector performance. The findings from the studies on private sector efficiency in the developed countries showed that improved efficiency is attributed to internal management and operational capability, level of competition for the market, scale of operation, economies of scale, market concentration and regulation. These factors were identified and used to explore the PSI situation in Ghana.

This study seeks to explore the role of these explanatory factors of private sector efficiency and service quality. In this thesis, the focus is on the firm level efficiency and service quality, and identifying the factors that explain the private sector performance in urban solid waste collection. The factors to explore include private sector capacity (scale of operation, skills/capabilities), and regulation (instruments, competition, contract obligations, political interference) and their implications for policy and sustainable urban solid waste management. Chapter 5 presents the evolving private sector involvement over the past two decades and households' involvement, while chapter 6 presents the level of private sector performance in terms of productivity and service quality. Chapter 7 presents the influence of private sector capacity on performance, while chapter 8 presents the influence of regulation on private sector performance.

4   Research Methodology

4.1   Introduction

This chapter aims to build on previous chapters and to provide the methodology for the study. This study seeks to explore the internal and external factors influencing private sector performance. To explain the differences in performances between private sector organisations, a framework used for the study combines three concepts to explain the variations in performance. These concepts are capacity (physical and operations management capacity), regulation (contractual relations and regulatory mechanisms) and the evolving processes of involving private sector and households in service provision.

    The chapter is organised around five sections. The next section explains the research strategy and how variables were measured and operationalised. Section 4.3 explains the study area, unit of analysis, sample size, sampling procedure and data collection methods and analysis. It also explains the reliability and validity of the measurement scale used for the various explanatory variables. Finally, section 4.4 summarises the methodology for the study.

4.2   Research Strategy and Operationalisation of Variables

Research Strategy

Research strategy for the study involves the use of both qualitative and quantitative data collection using a survey method. Sources of data will be key informant interviews, questionnaire survey and secondary data. Sources of quantitative and qualitative data can be both internal and external to the organization. A combination of qualitative and quantitative data is important, since by combining qualitative with quantitative information, a deeper understanding of the organization can be achieved (Lusthaus et al., 1995).

Assessing Performance

Table 4.1 presents the operationalisation of the variables and indicators for measuring performance as well as the definitions of concepts and indicators to be measured. The choice of indicators for efficiency and service quality assessment were guided by three considerations: (a) indicators from the literature review; (b) availability of performance data (c) going beyond exploratory study to empirical study by measuring key performance measures common to private companies.

Productivity and efficiency
Productivity is defined as the ratio of outcome (or output) to input resources used. For example the ratio of waste collected in tonnes to a vehicle, where output is the waste collected and input is the vehicle used. Productivity is also defined as productive efficiency because it results in cost savings that directly contribute to profitability. The major challenge posing the study of private firms' performance is the acquisition of financial viability data such as return on investment, return on sales, and profits. Private firms were reluctant to provide these financial data and therefore analysis of economic performance is difficult. Most of the companies surveyed for this study did not provide such data.

Quality of service

Service effectiveness is defined as the level of outcome accomplished, for example, the level of satisfaction that customers experience with services, the number of waste pickups made, the amount of revenue collected from customers, and the number of customers that are served. In this study, quality of service is used as a measure of performance of solid waste service. The measures of quality of service used in this study are reliability of service, customer satisfaction, responsiveness to customers, and environmental cleanliness.

Table 4.1 Operationalisation of Performance Variables

Variables	Definition	How indicators are measured
Productivity	Productivity is defined as the ratio of output to input. It is a measure of efficiency.	o Vehicle productivity – tonne of waste collected daily per vehicle (according to vehicle type and size), o Vehicle utilization – ratio of actual waste collected to company capacity, o DEA technical Efficiency – measures firm capability and management o DEA scale efficiency – measures scale of operations
Quality of service	Quality of service is defined as the extent to which the objectives of the service are met in practice. It is a measure of effectiveness.	o Reliability of service – regular waste collection and on specify days, and frequency of service, o Customer satisfaction – about littering and waste piles and overflows o Sanitary conditions and waste overflow – environmental cleanliness, cleanliness around communal containers, cleanliness of service areas, o Response to complaints of service delivery,

Assessing explanatory factors of performance

Table 4.2 presents the operationalisation of the variables for measuring the internal and external factors influencing performance. The definitions of concepts, variables and indicators to be measured are presented.

Scale of operation
This refers to the physical capacity of the companies. It is measured using the number of company vehicles (as proxy for company size), market share of waste collected, and the vehicle conditions.

Skills and capability
Service operations are intended to ensure that customers are served and the service objectives are met. The operation management capacity seeks to measure capacity for vehicle operations planning, operations supervision, and maintenance management, and quality of operation management employees. In this study, the measures of skills and capability for operations management used are operations planning capacity, maintenance management capacity, span of supervision, and quality of employees. The operations planning capacities involves the ability of the operations manager to plan waste collection and establish rational vehicle routes, route plans and vehicle logs for the service operations. The maintenance management capability is the ability of the technical personnel to plan in-house maintenance, prepare maintenance schedule to prevent equipment breakdown and long downtime duration. The operations supervision involves the ability of technical personnel to monitor collection services and emergencies.

Assessing external factors of regulatory mechanisms and contract arrangement

The regulatory mechanisms refer to the mechanisms for accessing information, controlling user charges and quality of service rendered. In this study, the contractual relation is defined as any contract and cooperative arrangement between private enterprise (formal) and the government (local and central), in which there are arrangements for mutual benefit, and each party adheres to obligations.

Table 4.2 Operationalisation of internal and external factors explaining performance

Variables	Definition	How variables are measured
Company capability	Service operations are intended to ensure that customers are served and the service objectives are met.	o Operations planning and scheduling – Route/rational vehicle routes, vehicle logs, pick-up scheduling, o Operations supervision – ability of technical personnel, o Vehicles maintenance capacity/schedule – In-house maintenance, maintenance schedule available,
Scale of operations	This refers to the physical capacity of the company.	o Number of company vehicles, o Market share of waste collected – ratio of quantity of waste collected by a company to the total waste collected in the city
Regulatory arrangements	This refers to the mechanisms for accessing information, controlling user charges and quality of service rendered, and roles and obligations in partnership.	o Regulations of access to information – frequency of report on operations, frequency of monitoring o Regulation of user charges – setting of charges, frequency of tariff review, levels of unit price per tonne of waste collection, and user charges, share of revenue from users and subsidy levels, o Regulation of quality of service – performance targets, targets complied, adherence to regulations, use of contracts o Regulatory capacity and autonomy – Payment for services and subsidies, sanctions applied to defaulters, regulatory framework complied, autonomy to apply policy and enforce laws, o Contractual obligations – adherence to conditions of contract, o Contract mechanism – duration and competition for contract, contract agreement signed, permission of entry with agreement, o Service mix – service and franchise contracts mix, with or with subsidy from local government/WMD,

| Private sector and Households' involvement | This refers to the evolving processes of involving private sector and households in service provision for better service quality and increase coverage. | o Zoning of service areas and zonal monopoly – single or multiple actors serving a zone, size of delineated zone or service area
o Trend in service coverage/level
o Customer education – by public and private sectors
o Pay service charges by households
o Households complaints – prompt response to complaints by companies
o Monitor service quality – households involved in service quality monitoring, customer scored card system,
o Waste recycling and minimization practices |

Private sector and households' involvement refers to the evolving practice of involving private sector and households in service provision for better service quality and increase coverage. The dimensions for assessment include the zoning of city into collection zone, cost recovery, paying for services, customer orientation, and waste minimization.

4.3 The Study Areas, Data Collection Instruments and Data Analysis

4.3.1 The Study Areas and Unit of Analysis

Figure 4.1 presents the map of Ghana showing the location of the study cities. The five largest cities (Accra, Kumasi, Tema, Takoradi, and Tamale) in Ghana were chosen for the study. These cities were chosen for the study because of the rapid urbanization and expansion of their municipal or metropolitan setting. According to the 2000 Population and Housing Census (Ghana Statistical Service, 2001), the urban centres of Accra, Tema, Kumasi and Sekondi-Takoradi accounted for about 44% of the total urban population, with some 350 urban centres accounting for the remaining 66%. As of 2000, Accra and Kumasi alone accounted for about 34% of the total urban population whiles the other three cities accounted for the remaining 10%.

The criteria for the selection were based on spatial distribution and the size of population (five largest cities in the country), and cities which benefited from World Bank's project of introduction of private sector participation (PSP) in solid waste management in the early 1990s. Kumasi is the second largest city in Ghana after the capital Accra. Geographically, Tamale is located in the northern part of Ghana, Kumasi is the centre, and Accra, Tema and Takoradi are along the coast. The population figures of the cities, number of private companies and sample size for this study are presented in Table 4.3.

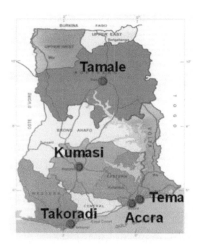

Figure 4.1 Map of Ghana showing location of the five study cities

There are 42 private companies involved in solid waste services in the five cities. The unit of analysis is the private service providers involved in solid waste collection. Survey method was used to collect data from private organisations involved in solid waste management service in five cities in Ghana (Accra, Tema, Takoradi, Kumasi, and Tamale). A sample of 25 private companies from the five cities was survey. For the analysis of productivity, secondary data on 33 companies in the five cities were used for the study. Data were also collected from five public organizations which are the five municipalities that have responsibility for waste management in the cities.

Table 4.3 Sampling of private solid waste companies in the five cities

City	2000 Census Population*	2007 Popu- lation*	Area (Km²)	Number of companies	Sample size
Accra	1,658,937	2,040,283	228	18	7
Kumasi	1,170,436	1,439,489	250	8	6
Tema	506,400	622,808	368	7	6
Takoradi	393,881	484,424	385	3	2
Tamale	275,436	338,752	750	6	5

*Excludes the transient population who come to do business and return

4.3.2   Data Collection Instruments

Company survey
In order to test the hypotheses and analyze the empirical relationships, a structured questionnaire was developed and pre-tested before it was used for data collection from companies in the cities. Both qualitative data on the determinants of performance and quantitative data on the performance were collected from each company. Secondary quantitative data on quantity of waste collected (disposal records) were collected from public organisations in the cities.

The survey questionnaires, measuring all the constructs for this study, were sent to 42 companies in the five cities but only 25 of the companies accepted to complete the questionnaire. The directors of the other companies refused to accept the questionnaire on the grounds that previous data from their companies were used against them. After follow ups to collect the completed questionnaires, 25 questionnaires were received from the managers, and this corresponds to an overall response rate of 56%. The key top management staff (managers and directors) of the companies completed the questionnaires.

Household survey
Data were collected from the residents using household surveys for the assessment of quality of service and mode of services. A random household survey of 1500 households was conducted in the five cities. Respondents' perceptions about the existing service quality (reliability, clean surroundings, accessibility, etc.) were rated on an ordinal scale from 1 to 5, with 1 being "strongly disagree" and 5 "strongly agree". Scores were calculated using the ratings to determine the quality of service for each respondent and company.

Key Informant Interview
Interview of local authority waste management directors and waste managers of private companies in all the selected companies was conducted to find out more about the contract arrangements, trust and cooperation in the contract relations, regulations, company's capacity, and the performance of the companies.

Secondary Data
Secondary data on the number and types of vehicles, quantities of waste collected, and number of trips made by each company were collected from the disposal sites and the Assemblies to cross check information given by the companies. The productivity of the companies was computed from these records.

4.3.3 Data Analysis

Data analysis consists of categorizing data, cross-tabulating and statistical analysis to assess performance and the variations in performance between well and poor performing companies. Comparative analyses of 25 private companies were conducted to assess performance and explore the key factors that drive performance. Comparative analysis was made between private companies involved in the solid waste service delivery within the

same city and across cities to assess the extent of performance and establish factors that explain the differences in the performance. A comparative
study of service providers was done to establish their relative performance
and the well performing companies. The testing of the hypothesis was
done by analysing the data statistically by using regression analysis, paired
t-test and correlation to determine the level of significance.

Test of hypotheses using regression and paired t-test analyses

Figure 4.2 presents the hypothetical relationships to be tested. The theoretical relationships between the explanatory variables and the performance variables were tested using paired t-test and regression analyses. The
theoretical hypotheses are tested in chapters 7 and 8 and subsequently interpreted and discussed. The significance of the empirical relationships between the variables was determined and the empirical results interpreted.

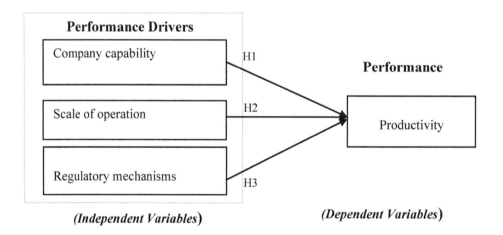

Figure 4.2 Hypothetical relationships to be tested

Dependent Variables

Table 4.4 presents the summary of dependent and independent variables for the analysis. For the purpose of this study, the dependent variable used to assess the performance of the companies was the productivity. The quality of service of the companies was not used for the analytical analysis due to the subjectivity of the customers' responses to the attributes.

Independent Variables

The variables that have effects on the performance of private firms are the independent variable. The independent variables were classified into three categories as discussed in the theoretical framework. These categories are the company skills/capability, scale of operation and regulation mechanism. The scale of operation was measured using the market share and size of the companies as proxy. The organizational size gives indication of the relative scope of operations as well as a measure of the physical resources for the solid waste service. The physical resources encompass the trucks and equipment for solid waste management. For skills/capability variables, a five-point measurement scale ranging from 1 (strongly disagree) to 5 (strongly agree) will be used to measure each survey item. For the regulation variables, a five-point scale ranging from 1 (very inadequate) to 5 (very adequate) was used to measure each of the variables. The reliability and validity of these scales was assessed through the analysis of internal consistency and the unidimensionality of each scale. For service quality variable, a five-point scale ranging from 1 (very poor) to 5 (very good) was used to measure each of the service quality attributes. The following formula will be used to calculate the service quality score.

$$Service\ Quality\ (\%) = \frac{\Sigma((n_1 \times 1) + (n_2 \times 2) + (n_3 \times 3) + (n_4 \times 4) + (n_5 \times 5))}{\Sigma(N \times 5)} \times 100$$

Where
n_1 = number of respondents rating service attribute as very poor (1)

n_2= number of respondents rating service attribute as poor (2)
n_3= number of respondents rating service attribute as fair (3)
n_4 = number of respondents rating service attribute as good (4)
n_5 = number of respondents rating service attribute as very good (5)
N = Total number of respondents

Table 4.4 Summary of variables for regression analysis

Theoretical concepts	Variables	Data source and scale of measurement
Performance	Productivity,	Ratio scale using waste quantities
	Quality of service	3-item scale from survey
Management capacity	Vehicle route planning	3-item scale from survey
	Operations scheduling	2-item scale from survey
	Supervisory capacity	2-item scale from survey
	Employees' moral	2-item scale from survey
	In-house maintenance capacity	Dummy, 1 for capacity, 0 elsewhere,
	Maintenance schedule	An item scale from survey
	Company size	Number of vehicles
	Company's experience	Number of years of service
Contract arrangement	Adherence to contract obligation	4-item scale from survey
	Contract mix	Dummy item, 1 if company operates two service, else 0,
Regulation mechanisms	Regulation of access to information	3-item scale from survey
	Regulation of tariff setting	2-item scale from survey
	Regulation of tariff level	2-item scale from survey
	Regulation of quality of service	4-item scale from survey
	Regulatory capacity	4-item scale from survey
	Financial autonomy to regulate	4-item scale from survey
	Managerial autonomy to regulate	2-item scale from survey

The items for measuring each of these variables and the results of reliability and unidimensionality analysis are presented in the Annex.

Reliability and Unidimensionality of Variables

A five-point scale was used to measure each item of the measurement scale. The reliability and unidimensionality (internal consistency) of the scale items were assessed using Cronbach's alpha and factor analysis respectively. Factor analysis is conducted as part of the analytical procedure to establish high-quality measures for the variables of skills and regulatory mechanisms. For the purpose of this study, two tests are used to assess the overall quality of the research design. The quality of a measure is perceived as the extent to which it possesses two desirable measurement properties: (1) reliability and (2) unidimensionality.

Reliability. To assess the reliability of a measure of variable, an internal consistency method called Cronbach's alpha (Cronbach, 1951) was used. This technique provides a unique estimate of reliability for the measure under analysis. The reliability of a measure concerns the extent to which it yields the same result on repeated trials (Sullivan and Feldman, 1994). Theoretically, alpha may vary between 0 and 1, corresponding to a completely unreliable measure and a perfectly reliable measure respectively. As a general rule, reliabilities should not be under 0.80 for widely used measures (Carmines and Zeller, 1994). At that level of Cronbach's alpha there is very little random measurement error. However, Cronbach's alpha of 0.6 is acceptable for scale measurement. The Cronbach's alpha for each of the variables is provided in the Annex.

Unidimensionality. Ideally, a scale should only measure a single underlying phenomenon. The unidimensionality of a scale can be assessed by means of factor analysis. Factor analysis consists of a variety of techniques for discovering clusters of interrelated variables or scale items. A factor is defined by a set of items that are more highly correlated with each other than with other items. For the present purposes, principal components analysis is the most suitable technique. For all the scales, the eigenvalues and the percentage of variance explained by the components are provided in the Annex.

4.4   Conclusion

This chapter explained the research strategy and the operationalisation of the variables. The explanatory factors of performance (capacity for service operations management, contractual relations, and regulation mechanisms) and performance variables (productivity and quality of service) were defined. The chapter explained study settings and the research methodology for data collection and analysis. The study was conducted in the five largest cities (Accra, Kumasi, Tema, Takoradi, and Tamale) in Ghana. These cities were chosen for the study because of their rapid urbanization and expansion. Qualitative and quantitative data were collected from five public organizations and twenty-five private companies. Secondary data on companies in the five cities were used for assessment of company productivity. A random household survey of 1500 households was conducted in the five cites. Households' perceptions about the existing service quality were rated on a scale from one to five and scores calculated. Household data were analyzed to determine the quality of service and residents' satisfaction. The implications of the study for performance improvement, regulatory policy, and sustainable service delivery have been explored. The statistical analysis and the interpretation of t-test and regression results are presented in chapters 6, 7 and 8 for the variables involved. The next chapter analyses and discusses the productivity and service quality of the companies.

5 Evolving Practices of Involving Private Sector and Households in Service Provision

5.1 Introduction

Chapter 2 reviewed the local context of legal, policy and regulatory frameworks and the historical developments of public and private provision of solid waste management. It was identified that the local governments (as called Assemblies in Ghana) have full constitutional responsibility for the management of solid-waste collection and disposal. However, the central government appoints chief executives to run local governments and therefore the chief executives may not have the free hand to manage solid waste, implement policies, and enforce legislation on solid waste due to politics. In addition, the central government controls the Assemblies through the Ministry of Local Government and Rural Development (MLGRD). The MLGRD formulates solid waste policies; oversees the role of the local governments; approves subsidies for provision of solid waste services; monitors the activities of local governments; approves the by-laws, budget allocation, and proposals to increase existing taxes; and passes directives as determined by state legislation. These controls may pose problems for involving the private sector and households in solid waste service provision and therefore service delivery may not be sustainable. It is believed that the central-local government relationship has implications for the governance of solid waste collection in the cities.

The central government policy towards private sector-led development for all sectors of the economy, led to the introduction of private sector provision of solid waste collection in Ghana. This chapter explores the evolving practices of involving private sector and households in solid waste service provision and identifies policy issues for service sustainability in terms of cost recovery and better service quality. It explores the different forms of private sector involvement (PSI) and how private sector has evolved over two decades (1990-2010) in solid waste collection as well as households' participation in solid waste management. The focus of the 'evolving involvement of private sector' is on how PSI was started and organized in the past and present, why changes in public and private arrangements (variant forms of PSI) happened, and what are the local conditions and context (policies, legal and capacity) necessary for better and sustainable service provision.

The dimensions used to explore the evolving involvement of private sector over two decades (1990-2010) are categorised into four: (1) zoning of cities into service areas for more private sector participation (comparison of population per collection zone), (2) evolving forms of public-private arrangements that exist in practice in the five cities (zonal monopoly, number of companies, nature of contract, companies' cost recovery arrangements – user charges and subsidies – and companies' market shares of waste collected), (3) trend in service coverage and level of service (proportion of house-to-house collection, proportion of communal collection, public collection, private collection, collection rate), and (4) implications of current practices for policy and regulation at the local level.

The analysis of households' involvement (awareness and participation) seeks to explore whether residents were involved in the service provision arrangements. The interactions among public-private-community are seen as a contributory factor to the success of service provision and promotion of willingness-to-pay for service. The level of household participation depends on the level of public awareness of solid waste-related issues (Wilson et al., 2001). The public support for solid waste management can be greatly increased if the public is fully and well-informed of the reasons behind the actions and the intended benefits. The analysis of households' involvement in this study covers awareness of solid waste-related public health issues, willingness-to-pay more for collection service, involvement in service quality monitoring, and willingness to support recycle and source-separation activities.

The chapter is organised into four sections. Section 2 presents the evolving forms of PSI in the five cities (Accra, Kumasi, Tema, Sekondi-Takoradi, and Tamale) and public-private arrangements in solid waste collection. The section looks at the contracting and cost recovery arrangements as well as the proportion of waste collected by each company. It also examines trend in service coverage (collection rate) which is the proportion of solid waste generated which is actually collected in the five cities. This section also examines the level of service (proportion of house-to-house and communal collection) and proportion of public and private collection. Section 3 examines the households' involvement in service provision and awareness of solid waste impact. The last section draws summary and conclusion from the analysis and to provide better understanding of private sector performance and regulation to be discussed in the subsequent chapters.

5.2   Evolving Involvement of Private Sector in Solid Waste Collection Service

This section examines and discusses how PSI evolved in the five study cities (Accra, Kumasi, Tema, Takoradi, and Tamale) after the German Agency for Technical Co-operation (GTZ) project in 1985 to build capacity of the public sector in solid waste collection. The evolving involvement of private sector is focused on when each of the companies was engaged for the first time and the form of private sector involvement with or without contract, and whether each company had a zonal monopoly, capacity, and standard technology for the service delivery.

5.2.1   Evolving Technologies and Service Levels

Prior to 1985, community incinerators called 'boiler' and locally called *'bola'* were in used in the urban centres. But all the *'bola'* had broken down by 1985 so the solid waste was dumped on the ground at all the *'bola'* locations. In 1985, Waste Management Departments (WMD) were established in the five cities selected for this study, equipped with vehicles for communal collection and funded by the German Agency for Technical Co-operation (Deutsche Gesellschaft fur Technische Zusammenarbeit GmbH – GTZ) prior to the establishment of the Local Government in 1988. The solid waste collection was the role for the then city council, but now the Waste Management Departments alone without PSI. With the assistance of a German company Gopa under the GTZ project in 1987, there was regular replacement of worn-out vehicles and equipment parts, and provision of funds for capital and recurrent costs.

There was ground evacuation of solid waste at all *bola* locations and central containers were used to replace the broken down *bola*. Animal drawn carts using donkeys were used to start house-to-house collection service in high income areas of Accra and waste collected was dumped into central containers. The house-to-house service involved 15 donkeys and 10 staff of the then city council. A portion of the waste from house-to-house service was used for composting. Each animal drawn cart could collects 3 to 4 trips a day with each trip covering collection from 25 houses. Despite the progress made under the GTZ project, services deteriorated after the project and public sector provision alone could not cope with the increasing urbanization of the cities.

There are two levels of services – communal and house-to-house service – rendered by the private sector. The communal service using central containers is provided in densely populated middle and low income areas. Householders participate in bringing their waste to these central communal containers. The collection crews using standard solid waste collection trucks (skip and roll-on trucks) bring empty containers to the small transfer stations and lift containers full of waste and transport to designated disposal sites. The technologies for communal solid waste collection have evolved from the use of conventional open trucks in the 90s to the use of standard solid waste collection trucks (skip and roll-on trucks).

House-to-house collection service is rendered in high income and middle income areas as well as for serving offices, hotels, shops, institutions, and industries where access roads permit the use of standard compactor trucks to pick waste directly at the doorstep of the premises. In areas inaccessible to compactor trucks, but communal collection service is rendered, certain informal cart pushers (using human powered push cart technology) collect waste from house-to-house for a fee and dispose of the waste into the central communal containers. The technologies for house-to-house solid waste collection have evolved from the use of low technologies (animal drawn cart, motorized tricycle, power tiller), to the use of conventional open trucks (Bedford truck, tipper truck, Tata pick-up, tractor with trailer), and now to the use of standard solid waste collection trucks (compactor and sideloader).

5.2.2 Zoning of cities into service areas for private sector involvement

Zoning in Accra

The number of sub-metropolitan districts (sub-metros) in Accra was increased to 6 in 2004 for administrative and public services delivery purposes by legislative instrument. The number of sub-metros was again increased to 14 in 2007 by further sub-dividing the original 6. For the purpose of engaging private companies in solid waste collection, Accra was delineated into 15 solid waste collection zones in 2007 using the boundaries of the sub-metro districts, and later on re-zoned in 2008 into 11 zones. Figure 5.1 is a map obtained from the Accra Waste Management Department (WMD) showing the collection zones and the companies in operation in 2007. This is an evidence of the practice of zoning and PSI in Accra.

The rationale of using the sub-metro boundaries is that the monitoring of the solid waste collection is done by the Sub-metro Districts Cleansing

Officers. However, the zoning along sub-metro boundaries does not allow mixing of low, middle and high income areas into a zone. Certain zones are completely high income areas whereas some are completely low income areas. The proper zoning of a city is a key factor for the success of private sector involvement in solid waste collection.

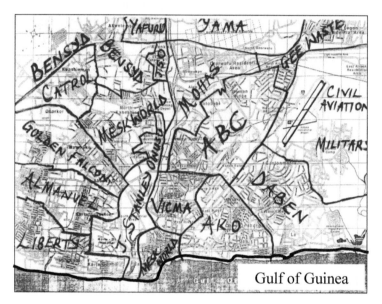

Figure 5.1 Map of Accra showing collection zones in 2007. Source: Waste Management Department of Accra Metropolitan Assembly.

Zoning in Kumasi

Four sub-metropolitan districts in Kumasi were zoned in 2000 into seven solid waste collection zones by the Waste Management Department (WMD). The zoning was intended to bring in more private sector companies and to enhance competition for the market. In 2007, the government re-categorised the four sub-metros into 10 by legislative instrument due to the increasing population especially in the peri-urban areas of the city. The WMD used the new sub-metros boundaries to re-zone the city into 10 solid waste collection zones. The zoning approach is the same as that of Accra, using sub-metro districts' boundaries. Figure 5.2 is a map showing the 10 sub-metros (collection zones) and the companies in operation in 2008.

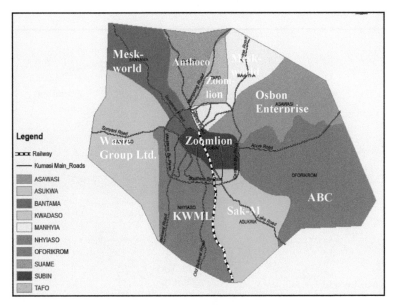

Figure 5.2 Solid waste collection zones and companies in 2008 in Kumasi.

Source of map of Kumasi Metropolitan Assembly: www.GhanaDistrict.com

Zoning in Tema

Four sub-metropolitan districts (sub-metros) in Tema were divided into six solid waste collection zones by the Waste Management Department and each zone assigned to a company. In 2007, one sub-metro (zone 3) was created as a new district, reducing the sub-metros from four to three and the solid waste collection zones to five. The city was divided into 23 communities (C1 to C23) when it was built for administrative and management purposes, but the zoning of the city was done by clustering adjacent communities to form zones. Tema is a model city with well planned streets making house-to-house collection appropriate. Figure 5.3 is a map showing the six collection zones and the companies in operation in 2006. Certain zones are completely high income group areas whereas certain zones are mixture of middle and low income groups.

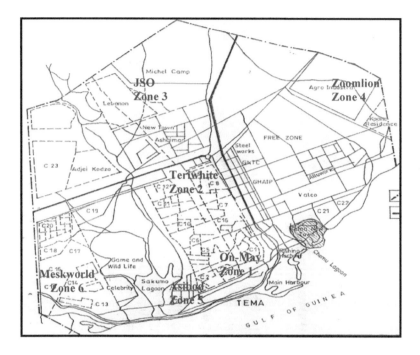

Figure 5.3 Map of Tema showing solid waste collection zones and companies in 2006
Source of map: Tema Metropolitan Assembly

Zoning in Takoradi

There were 3 sub-metropolitan districts in Takoradi in 2007. The boundaries of the collection zones were in line with the boundaries of the sub-metros. In 2009, the city was zoned into six collection zones by the WMD to allow for more private sector involvement. Figure 5.4 is a map showing the three sub-metros (collection zones) and the companies in operation in 2008.

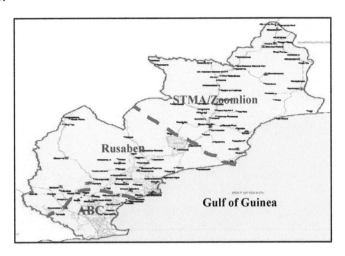

Figure 5.4 Map of Takoradi showing the solid waste collection zones in 2008
Source of map: www.ghanadistrict.com

Zoning in Tamale

There were 3 sub-metropolitan districts in Tamale in 2007 (Figure 5.5). The central part of the city is the central business district. The Assembly assigned some communities in one sub-metro to five small companies rendering house-to-house collection, while the Waste Management Department (WMD) of the Assembly and another private company were operating in two sub-metros and part of the sub-metro where the other five companies were operating. The sizing of the collection zone for the five companies was not based on any analysis. There was no map showing the collection zone of individual companies operating the house-to-house service.

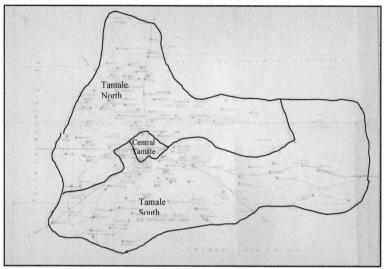

Figure 5.5 Map of Tamale showing three solid waste collection zones in 2007
Source of map: Tamale Metropolitan Assembly

Zoning of cities into service areas

One of the success factors of PSI in solid waste collection is the zoning of a city into solid waste collection zones and the granting of zonal monopoly to companies. This study explored to find out the practice of zoning of cities and whether zonal monopoly exists in the cities. Table 5.1 shows the size and number of collection zones across the cities. The size of collection zone in terms of average population of zones in 2008 varies from 88,800 for Tema to 185200 for Accra. A zone with a size of less than 50,000 people does not encourage economics of scale (Stevens, 1978). The size of collection zone (112,740 people) for five companies in Tamale was too small for economics of scale and this explain why they had low productivity (see chapter six). These five companies serve a total of 1275 households which for a proper zoning should be the number of households for a company. The number of households served by companies in other cities ranged from 500 to 7500.

In order to involve more private sector companies in solid waste collection, a service area is divided into collection zones of sizeable population (Cointreau and Coad, 2000; METAP, 2008). This provides opportunities for stimulating competition and better service provision, and also making it possible to match the capacity of the private sector. The size of each zone need to be determined for the capacity of four vehicles and also has to take into consideration cross-subsidization of user charges of low income residents from that of high income residents (METAP, 2008).

Table 5.1 Size of Collection zones across cities

City	Number of zones in 2000		Number of zones in 2008		Average Population per zone (2008)	Number of companies (2008)
	sub-metros	collection zones	sub-metros	collection zones		
Accra	6	15	14	11	185,189	18
Kumasi	4	7	10	10	143,723	8
Tema	4	7	3	7	88,833	7
Takoradi	3	3	3	3	161,221	3
Tamale	3	3	3	3	112,740	6

The considerations for zoning of the city are the level of services (communal and house-to-house) to be rendered and the income levels of the inhabitants. However, in all the cities, the zoning was done using sub-metropolitan administrative boundaries and therefore did not afford the cross-subsidization of user charges to support poor urban residents. The inappropriate zoning of the cities led to the creation of solid waste collection zones which are completely made up of low income areas and those which are completely high income areas. During competitive bidding no company wants to bid for zones which are completely low income areas. This is because the subsidy paid by the Assemblies for communal collection delays over a year. Besides, the willingness to pay in such areas is low and as such franchised house-to-house service does not work in such areas. This has brought competition among companies using their political influences to win high income zones. There was a clear case of court injunction in 2010 placed on the Accra local government (AMA) not to sign contract for a zone which is completely high income areas upon suit filed by one of the companies. The proper way is to demarcate those sub-metros which are completely high income areas and combine portions of high and low income areas into a zone for tendering. This will afford cross-subsidy and fairness to all companies working in the cities.

The companies enjoy zonal monopoly for house-to-house service delivery for the period given to operate in their respective zones, but not for communal collection in certain zones. This arrangement was evident in four cities. The zoning was to provide opportunities for more companies to be engaged and to bring about competition. The zonal monopoly for communal solid waste collection was changed to allow a second company to operate in the same zone where there was declining capacity of the first company that won the contract. This means that a company may operates zonal monopoly for house-to-house service delivery but serves communal collection in the same zone with another company. The second company is required to place its own communal containers at the same sites of the first company and pick them as and when they get full. This arrangement was used in zones where there was increasing solid waste generation without corresponding increase in the capacity of the first company to handle the increasing demand.

5.2.3   Evolving Trend and Forms of Private Sector Involvement

Evolving PSI in Accra

In Accra, service coverage, as a result of GTZ assistance in 1987, which had increased to about 80%, began to fall again. By early 1990, it had fallen to 45% because of limited financial resources for maintaining the trucks (AMA/Colan Consult, 1995). This decline necessitated the involvement of the private sector in solid waste collection at the time when there was a global move towards private-led provision of public services.

Table 5.2 shows the stages and the evolving involvement of private sector in Solid Waste Collection in Accra. Private sector companies started operation in Accra in the early 1990s with five private companies. These companies rendered franchised house-to-house service using low technology vehicles and collected user fees. These companies were given contract letters and maps showing collection zones by the then head of Accra City Council, now Accra Metropolitan Assembly. The Waste Management Department was involved in solid waste collection operation from 1985 with a gradual decrease in waste collection until 1999 when the number of companies increased from 5 to 11.

Table 5.2 Evolving PSI in Solid Waste Collection in Accra

Stages of PSI in Accra	Form of PSI
o In 1991 first company was engaged using donkeys,	o First franchise house-to-house arrangement with the experiences from the GTZ project using donkeys,
o In 1995 five companies were engaged,	o Franchise contract letters and maps were given to the companies. Received no subsidy from the Assembly,
o In 1999 an international-local joint venture company was engaged,	o Monopoly contract of the entire city, and the company sub-contracted the collection to 11 local companies,
o In 2001 seventeen companies were contracted to collect waste in the city,	o Six-month contracts covering both communal and house-to-house collection were signed,
o In 2006, a new company entered into the waste market,	o Company had a signed contract with Assembly but did not undergo competitive bidding,
o In 2010 twelve companies contracted to collect waste in the entire city involving 10 zones,	o Competitive tendering in 2008, but engagement with 3 months probation before contract is signed subject to satisfactory performance of the companies commenced in 2010,

Source: Field data collected from company survey and Key Informant Interview by author

In 1999, the collection service was contracted out (city monopoly contract) to an international-local joint venture company (called Ghana Waste Limited). The 11 private companies in operation before 1998 were engaged as sub-contractors by the main company (Ghana Waste). There was a World Bank project (Urban Environmental Sanitation Project – UESP) in Ghana in 1998 to introduce private sector in solid waste collection in five cities, but Accra was omitted from UESP because of the city monopoly contract where the whole city was given to Ghana Waste in 1999 to collect solid waste. The reason for the city monopoly was to source for foreign loan to procure solid waste collection vehicles and also to reduce the role of the Assembly in direct delivery of solid waste collection. The monopoly company charged the government US$12/tonne and paid US$4/tonne to the subcontracted local companies (Boakye, 2005). The city monopoly contract was terminated in 2001 by a new government and a six months contract was signed with local private companies, to be renewed every six-month subject to satisfactory service quality. The reason for the termina-tion of city monopoly contract was mainly the high cost of solid waste col-lection according to the new government. The government asked for price reduction but the company did not agreed to the request.

The practice of zoning of a city and zonal monopoly for solid waste collection was restored in 2001 after terminating the city monopoly con-tract. The contract bidding was not competitive but mimics selective bid-ding where the requirements for service provision from the companies were submitted to the Waste Management Department and evaluated be-fore companies were assigned to zones. The requirements submitted for evaluation by WMD include: (1) the company profile and experience of company in the waste collection business, (2) office location of the com-pany, (3) human resource level of the company and (4) equipment holding of the company: a company is expected to have at least five vehicles to en-sure economics of scale and flexibility in operation in case of breakdown of one vehicle. Each of the local companies was given a contract letter and a map of the area to operate. The collection vehicles were shared among the local companies and arrangement was made to deduct the cost of the vehicles over a period of time from payments to the companies. From 2001 to 2008 permit notes (letters) were given to the companies from time to time indicating the conditions and terms of contract (areas to operate, scope of services, price per tonne for lifting container and user charges for house-to-house service, and mode of payment). No contract was signed with the companies after the expiry of the first 6-months contract since the Assembly wanted to reduce the number of companies. Besides, the Assembly also owed the companies for over one year waste collected. There was a signed contract for a new company that entered the market in 2006, even though other companies had no contract from 2001 up to that date.

The number of companies increased from 11 to 18 by 2007, even though Accra was delineated into 15 collection zones in 2007. Three of the companies had no specific zones but were assigned to lift some communal containers from zones of other companies which were having difficulties with capacity to cope with the increasing volumes of waste, especially from markets and low income areas. This means that for zonal monopoly for communal collection to work, companies must have the capacity to cope with the increasing demand in the assigned zone. The capacities of the 18 companies vary with the size of collection zones and the quantities of waste generated.

There was a competitive bidding and bid evaluation in 2008 for ten contracts (one for each of the 10 zones). Figure 5.6 is a map of Accra obtained from Accra WMD showing the contract zones for the competitive bidding in 2008 and companies that won the contracts. An advert on the competitive bidding was placed in the print media in 2008 and some 30 bids were submitted by companies (including companies operating in other cities). In order to meet the contract requirements in terms of capacity (number of collection vehicles and personnel) 3 companies entered into joint venture to win a zone. One company had two zones and the other eight companies had one zone each. Seven of the companies exited from service provision. Some of them did not win contracts while others did not put in tenders for solid waste collection when request for tenders was made by Waste Management Department of the Assembly since these companies were not interested to continue with the waste collection business.

The public sector (Assembly) began to give more weight to the private sector involvement through the competitive bidding for ten new contracts and selection of 12 companies in 2008 but award of contract began in June 2010 with 3-months probation. The delay in the signing of contracts was due to change in government in January 2009 through national election.

This means that the Assembly is now managing ten signed contracts instead of the permit arrangements with each of the 18 companies.

**Figure 5.6 Map of Accra showing collection zones in 2008 for new contracts.
Source: Waste Management Department of Accra Metropolitan Assembly.**

The new contracts combined house-to-house and communal collection
services into franchised contracts to be signed in 2010. The communal service involving skip containers is rendered in the markets and transport terminals as part of the contract. The house-to-house collection where user
charging covers full cost without subsidy from the Assembly is rendered in
low, middle and high income communities. The companies have the responsibilities to: (1) register all households and give customer card, and

supply 60litre, 80litre, 120litre, and 240 litre bins to all households, (2) ensure total collection of solid waste generated at the zones of operation, (3) collect service charges of GH¢3.30 (US$2.5) from low income households, GH¢9.00 for middle income, and GH¢14.00 from high income households, and (4) companies are to pay GH¢12.00 per tonne for the cost of disposal from the service charges. There is a shift from subsidy payment for solid waste collection to the use of service charges from all households. This is a step to involve all households and make companies responsive to service recipients in the provision of better service quality. The funding of solid waste collection through subsidy from the central government and Assembly's common fund is not sustainable and does not ensure better service quality.

Evolving PSI in Kumasi

Table 5.3 shows the evolving stages of private sector involvement in solid waste collection in Kumasi. PSI started in the mid 1990s with one company, upon request by the company to provide service. The company was given permission to collect solid waste but had no contract with the Assembly and no subsidy was paid. The households' subscription was mainly voluntary. The company recovered cost from the user charges.

In August 1998, a greater part of the city (6 out of 7 zones) was given to another company and two contracts were signed (one for lease of equipment and the other for service delivery). There was no competitive bidding for the city monopoly contract in Kumasi. It was at the time when Ghana Waste was operating in the whole of Accra under city monopoly contract, and the government at that time did the same in Kumasi. The reason for the city monopoly was that the government wanted the Assembly to withdraw from direct delivery of solid waste services and play supervisory, monitoring and regulatory roles. The company signed contract with the Assembly for collection of communal and house-to-house waste over a 5-year period. In the lease contract, the solid waste collection vehicles belonging to the Assembly were leased to the company and a lease fee of

10% of service contract sum was deducted from the payments to the company to cover the cost of the Assembly's vehicles. During this period the first company was still in operation in the zone assigned by the Assembly. This was because the company was rendering house-to-house service and collecting user charges without subsidy from the Assembly. This reduced the financial burden on the Assembly, and that is what the Assembly has to work towards.

Table 5.3 Evolving PSI in Solid Waste Collection in Kumasi

Stages of PSI in Kumasi	Forms of PSI
o In mid 1990, first company started operation in Kumasi at Ayiga Extension.	o As first PSI, it was intended to operate franchise house-to-house collection, with user charges covering full cost. There was not contract with Assembly and no subsidy.
o In August 1998, greater part of the city was given to the second local company and a contract was signed.	o First monopoly contract in the city but no competitive tendering. It was a politically led private sector involvement, promoting monopoly instead of competition,
o In 2001, the third company was engaged and a second formal contract signed under the UESP World Bank Loan project.	o UESP was intended to experiment cost recovery through franchise contract for house-to-house collection, with user charges and a declining subsidy to cover full cost.
o In 2004, three new companies entered the waste market and were given letters (permission) to operate in Kumasi but no contract was signed.	o Reason for the permit was to reduce financial burden of the Assembly through franchise house-to-house collection with user charges covering full cost, and also more resources were needed.
o In 2006, a new company entered into the waste market,	o Company did not undergo competitive bidding and had signed contract with Assembly.
o From 1st January 2008, implementation of the city-wide-fee through pay-as-you-throw (PAYT) mechanism started and 10 contracts were signed – one contract for each and 8 companies contracted in the city.	o Collection contract covers both communal and house-to-house service. Reason for this initiative was to reduce financial burden (subsidy) of the Assembly and make private sector more responsible for cost recovery, and to introduce competition.

Source: Field data collected from company survey and Key Informant Interview

Another company was engaged in 2001 under the UESP to serve two zones (out of the seven zones). This was the second formal franchise contract signed and operations started in 2001. The project was designed to provide lessons for cost recovery of solid waste collection and therefore declining subsidies of 75%, 50%, 25% and 0% over a four year period were introduced and financed by loan facility from World Bank. The company also signed service contract for communal collection (of about 35% of its total collections) in the low income communities for which costs were fully covered through subsidy from the Assembly and Government through the Ministry of Local Government and Rural Development (MLGRD).

In 2004, three new companies were given letters (permission) to operate in Kumasi but no contracts were signed. The capacity of the company, which had the city monopoly contract, was declining and therefore these companies took advantage of the situation to provide service. These companies entered into the waste business to gain experience and establish relationships with the Assembly. By 2006, there were seven private companies involved in the solid waste collection and disposal, with varying capacity (number of equipments owned) and operating under different arrangements.

The implementation of the city-wide-fee to increase cost recovery started in 2008. There were 10 zones but 8 companies were involved in SWC since two companies had two contracts each signed with the Assembly. During the call for proposal, only few companies submitted proposals, so they were all engaged to operate in Kumasi. Unlike Kumasi, Accra had more bids for competitive bidding due to the increasing number of new companies operating in the three Assemblies sharing boundaries with Accra Assembly. The new entry into solid waste market in Kumasi is slow compared to Accra and its surrounding Assemblies. Under this new arrangement, where both communal collection and kerbside collection (house-to-house service) are rendered in the same zone, and the two are put together as one contract and signed. Thus, one contract covers both services if they are rendered in the same zone by one company. The provision of the two services in a zone was due to the zoning where a particular zone may consist of low income areas and middle or high income areas.

Recent issues of poor performance of two companies have led to their replacement by another in 2009. There was heaps of solid waste revealing that these companies were not picking up the waste regularly. Skip containers were full and the people had to dump the waste on the ground at the container sites. A contract signed between a company and Assembly is monitored by means of collection inspection. For the house-to-house service the pick-up of waste was monitored using tour schedule of collection provided by the company. For the communal collection monitoring, the frequency of pick-ups and overflow of waste onto the ground were monitored. According to a clause in the contract, any non-compliance is sanctioned and deduction of money for non-compliance is made when payment is made. Payment Certificate is prepared at the end of each month, but payments to the companies delay.

Evolving PSI in Tema

Private sector companies started operation in Tema in the early 1990s due to the planned nature of the city and the income levels of the residents. Five companies were in operation by mid 1990s rendering house-to-house collection service and collecting users charges without subsidy. They were engaged for 3 months contract and renewed afterwards. They were using tipper trucks but the Assembly requested them to use side loaders. In 1995 the Assembly reviewed its policy and requested all solid waste companies to phase out the use of open tipper trucks from the system and use standard vehicles (side loaders and compactors) for solid waste collection.

In 1997, there was a competitive bidding, under the UESP World Bank Project, for the selection of a contractor to render franchise house-to-house contract with declining subsidy. One of the five companies was selected and engaged after the competitive bidding to render the service. This was done in order to generate competition among the prospective private sector companies with the view to getting the best company at the lowest possible price. A private firm was awarded a contract to a zone depending on its capacity and the responsiveness to the bid. A basic requirement to qualify as a solid waste company, at the time, was to own at least two trucks with some previous knowledge in waste management or similar works as an advantage.

Table 5.4 shows the contracts, service characteristics and payment arrangements for engaging private sector in 2006. By 2006 there were 6 companies serving strictly in their zones, except one company with political connection which unofficially extended its house-to-house service to zones of other companies.

Table 5.4 Contracts and arrangements for payments in Tema in 2006

Company	Level of Service	Contract and Payment Arrangements
C15	Communal and House-to-house services	Communal service contract in 2003 (Subsidy from Assembly) and service contract for house-to-house with subsidy from Assembly
C16	Communal and House-to-house services	Communal service contract in 2003 (Subsidy from Assembly) and franchise house-to-house without subsidy from Assembly
C17	Communal and House-to-house services	Communal service contract in 2006 (Subsidy from Assembly) and franchise House-to-house without subsidy from Assembly
C13	House to house	Service contract in 2002 for house-to-house with subsidy from Assembly
C14	House to house	Franchise contract in 2000 for house-to-house without subsidy from Assembly
C18	House to house	Franchise contract in 2002 for house-to-house without subsidy from Assembly

Source: Field data collected from company survey and Key Informant Interview by author

The Assembly had both service contract and franchise contract arrangements with the companies. The service contract is signed for the communal service, which is rendered in the low income communities (Sakomomo, Kpong, and Tema Newtown), markets and transport terminals, and in areas where house-to-house service is rendered (communities 1, 2,

4, 7, part of 5) but it is difficult for companies to collect user charges. Two companies (C13 and C15) were rendering franchise (house-to-house) contract but requested for it to be changed into service contract, because the collection of service charges was difficult for them as the people were not paying. For the service contract, the revenue collection is done by the Assembly. Companies that shift the task of collecting user charges to Assembly may not be responsive to users and may not provide better service. The Assembly has 25 revenue collectors and pays 20% commission to the revenue collectors. The monthly revenue collection efficiency for 2009 ranged from 35% to 82%.

The franchise contract is signed for house-to-house collection where user charging covers full cost without subsidy from the Assembly. For the house-to-house service, residents pay for the cost of the bins, so the handling of the bins by the collection crew to avoid damage is a concern to the residents. There is also the tendency for the bins to be stolen, so the bins are kept in the houses with fence walls and sent out (set-out) to the road kerbside on the collection days.

Evolving PSI in Takoradi

There are currently three (3) private solid waste companies operating in the twin city of Sekondi-Takoradi together with the Waste Management Department. The first company (C31) engaged by the Assembly started operation in 2000, followed by the second company (C32) in 2005 and the third (C33) in 2007. Each company has a zonal monopoly to operate house-to-house collection in a delineated zone, except one company which unofficially extended its house-to-house collection to zones of other companies. But for communal collection one company was allowed by the Assembly to collect containers from zones of other companies. The Assembly (STMA) also rendered about 50% of the service for both house-to-house and communal container collection prior to 2007.

The first company was operating in the city prior to the World Bank (UESP) project to introduce private sector involvement in SWC. Under the

UESP the first company signed a contract with the Assembly in August 2000 to render both house-to-house and communal container collection. The contract under the UESP ended in 2005 and the company was given opportunity to render service after the UESP without a contract with the Assembly, but a form of arrangement in writing. The company renders twice and thrice a week house-to-house service depending on the generation rate and payment arrangement with the Assembly.

The second company was engaged to render solid waste collection in October 2006 with five years contract with STMA, to render both house-to-house and communal container collection. It has a zonal monopoly to operate in a delineated zone and provides twice a week house-to-house service. No provision was made in the contract for payment of subsidy for house-to-house collection, but in 2009 the Assembly requested for detail accounts of revenue received through house-to-house user charges for consideration for subsidy. The company collects user charges (revenue) from about 1800 customers being served under the house-to-house service, out of about 3000 potential customers in the service zone. Most of the customers opted out of the service as they were not willing to pay due to the poor service quality.

Company C33 started operations in 2007 with a contract signed with the Assembly. There was arrangement between the Assembly and the company for provision of communal collection service, but there was no competitive bidding. In 2009, the city was re-zoned from three into six zones and a map prepared. This was intended to allow more competition and more companies to operate in the city. The rationale for the re-zoning was partly due to the dwindling capacity of the first two companies, due to vehicle breakdowns and old age of vehicles, and partly due to increasing population and expansion of the city. The re-zoning is also to afford the opportunity for competition and increase involvement of more companies.

Evolving PSI in Tamale

The PSI in Tamale started with one company (Sava Waste) during the UESP-I from 2002 to 2004 where a franchise contract with a declining subsidy was signed. In 2005, as part of UESP-II, a call for proposal (tenders) from private companies to provide house-to-house solid waste collection service was made through the media (including radio announcement). Sava Waste did not submit tender due to loss of interest as the company was not paid as scheduled, and as of 2009 an amount of GH¢30000 was not paid to the company. Out of 43 private companies that submitted both financial and technical proposals, five companies were selected.

A franchise contract with subsidy was supposed to be signed with the each company, but no contract or agreement was signed with each of the companies. A letter of introduction was given to the companies to start operations. The five private companies providing house-to-house service were given part of one out of the three sub-metro zones in the city. The companies were asked at a meeting to submit monthly invoice for payment. The cost of the franchise house-to-house collection service was recovered through user charging and subsidy from the Assembly/government. The franchise arrangements for the provision of house-to-house collection were terminated at a meeting in July 2008. The termination of the arrangements by the Assembly was due to financial constraints and delay in payment. Three months invoices out of twelve months submitted by the companies for payment for solid waste collected were not paid as of June 2009, one year after the termination of the unsigned contract.

A new company entered the solid waste market in 2006 and was assigned one sub-metro (the central business district) to serve. The provision of communal service by the new company in areas where the Assembly was operating was due to the limited capacity of the Assembly. The residents did not pay for the communal service, so the payment for the service contract was through subsidy from the Assembly and central government. The new company took over the areas of operation by the other private companies after the termination of the arrangements and gave free bins to the residents and collected higher user charges from the people served, but some people opted out due to the high user charges.

Comparative analysis of Trend and Forms of PSI

PSI started in Accra and Tema, and later extended to the other cities. Figure 5.7 shows the trend in the cumulative number of companies from 1990 to 2010 across the cities. Private sector companies started operation in Accra and Tema in the early 1990s whereas the other cities started in the late 1990s. There was a sharp increase in the number of companies across the cities between 2002 and 2007. The companies in Accra and Tema increased to 18 and 6 respectively by 2006.

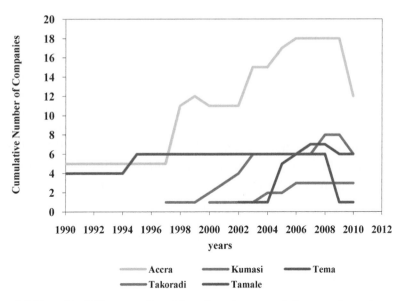

Figure 5.7 Trend in PSI across the cities, showing cumulative number of companies from 1990 to 2008. *Source: Field survey*

With the recent entry of a new company with high capacity (trucks, human resources and finance) in 2006 and competitive bidding in Kumasi (2007) and Accra (2008), the small companies with limited financial resource and trucks are exiting from the waste market. This is shown in figure 5.7 by the decline in the trend of the cumulative number of companies in the cities.

The small companies are exiting from the waste business due to the increasing competition and the capacity requirement for bidding a contract. For instance, some companies in Accra (5 companies), Kumasi (3 companies) and Tamale (5 companies) have exited from the solid waste market. As part of the bidding process, the number and condition of solid waste trucks are inspected by staff of the Assembly involved in the procurement of companies for the solid waste services.

The forms of PSI were franchise without subsidy, franchise with declining subsidy, service contract, and service permit without contract. Table 5.5 presents the scope of services and contracts with and without subsidies in 2007. The results revealed that there were financial arrangements in place for paying the private companies. Eight companies rendered house-to-house service with subsidy and 16 companies rendered house-to-house service without subsidy. Three companies in Kumasi were rendering house-to-house collection service without subsidy. The house-to-house collection service with declining subsidy was introduced as part of the Urban Environmental Sanitation Project (UESP) in Tema in 1997 and in Kumasi, Takoradi and Tamale from 2000 to 2005 to stimulate the introduction of user charging and to increase cost recovery. Those who did not receive the subsidy for house-to-house collection are private companies who have signed contract or have been given the permission to collect waste in particular areas but did not have any contract with the Assemblies. The user fees for full cost recovery of house-to-house collection without subsidy are approved by the Assemblies.

Table 5.5 Scope of services and contracts with/without subsidies in 2007

Cities	Companies	Scope of services				Contract signed/Year
		House-to-House without subsidy	House-to-House with subsidy	Communal collection with subsidy	Commercial/industrial collection	
Accra	C6	x		x	x	No
	C5	x		x		No
	C8	x		x	x	Yes (2006)
	C2	x				No
	C9	x		x		No
	C3	x		x		No
	C4	x		x	x	No
Kumasi	C20	x		x	x	No
	C24	x				No
	C21	x		x	x	Yes (2006)
	C22		x	x	x	Yes (2002)
Tema	C15	x		x	x	Yes (2003)
	C14	x		x	x	Yes (2000)
	C16	x		x	x	Yes (2003)
	C17	x		x	x	Yes (2006)
	C34	x		x	x	Yes (2003)
Takoradi	C31		x	x	x	Yes (2005)
	C32		x	x	x	Yes (2006)
	C33	x		x	x	Yes (2006)
Tamale	C26		x			No
	C27		x			No
	C28		x			No
	C29		x			No
	C30		x			No
	C25	x		x	x	Yes (2006)

X - Services with/without subsidy rendered by companies

In the early 2000, more companies entered into the solid waste market in Accra and the other cities, and contracts were signed with some of the companies in Accra (6 months contract), and five years contract in Kumasi, Tema and Takoradi. Apart from four contracts, one in each city, under the Urban Environmental Sanitation Project which underwent competitive bidding, the rest did not go through competitive bidding. The competitive bidding began in 2007 in Kumasi and Accra, a shift from non-

competitive bidding and issuing of permit letters to a more formalized contracting-out procedure. This is because there is pressure on the Assemblies
by the association of solid waste companies (called Environmental Service
Providers Association) demanding signed contracts and regular upward
review of service charges and collection fees to reflect increasing inflation
and foreign exchange rates. Besides, the cost of solid waste covered by
subsidy is now a burden on the Assemblies, resulting in long delay in
payments.

The payment of subsidy by the Assemblies and central government is
not sustainable and does not encourage competition and signing of contract. The fear of default in payment by the Assemblies accounts for the
unsigned contracts. The companies provide services with different financial arrangements. A franchised contract (signed or unsigned) for house-to-
house service was provided by the companies with or without subsidy from
the Assembly. The cost recovery for the franchise house-to-house service
provided under the Urban Environmental Sanitation Project was user
charging with declining subsidy from the Assembly and the project. The
arrangement for communal service is a service contract with subsidy from
the Assembly and central government with or without user charging.

5.2.4   Trend in Solid Waste Collection Coverage and Service Level

Accra Coverage and Service Levels

An analysis of waste generation and coverage was conducted using data
from secondary records on waste collection and disposal. The coverage of
solid waste collection is the ratio of the population served in an area to the
total population in the area. Collection coverage can be interchanged with
collection rate which is the ratio of the total amount of waste collected
from an area to the total amount of waste generated in that area or community. Waste collection rate was determined to assess the contributions of
the private companies involved. Research on solid waste generation in Accra and Kumasi estimated the household waste generation rates to be in the

Evolving Practices of Involving Private Sector and Households 155

range of 0.35kg to 0.75kg (Saleh, 2002; Owusu-Ansah, 2008). For the purpose of comparing all the five cities a waste generation rate of 0.6kg/capita/day was used to calculate the amount of waste generated for the various years. The assumption is that this generation rate make up for non-domestic waste and waste generated by transient population at market and public places.

Figure 5.8 shows the trend in solid waste collection rates, service level, and private sector involvement for Accra from 1990 to 2009. The annual amount of waste collected in the Accra metropolis increased from 405908 tonnes in 2004 to as high as 601268 tonnes in 2006. The average daily waste collected in 2007 was 1630 tonnes. In Accra, the waste collection rate increased from about 67% in 2005 to as high as about 93% in 2006. This increased in waste collection rate in the year 2006 could be attributed to the preparation towards the national independence jubilee in 2007. The Government increased its budget for environmental sanitation in the city. The increase in waste collection in the year 2006 could also be due to the market entry of a new company. According to officials of the WMD the government paid companies part of monies owned them to indirectly increase collection.

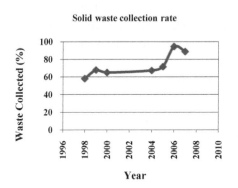

Solid waste collection rate

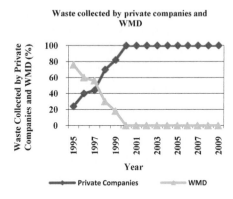

Waste collected by private companies and WMD

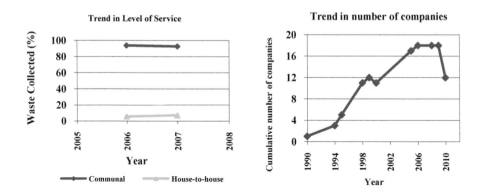

Figure 5.8 Trend in Solid waste collection rates for Accra and Private sector in-volvement. *Source: Analysis of data collected from Accra Waste Management Department*

In 2006 and 2007, communal system of waste collection was 94% and 93% of the total waste disposed of by all the companies as compared to 6% and 7% for house-to-house collection. This shows there is reliance on the communal system of collection with huge subsidy associated with the system. This also accounts for the inability of the government to pay the companies regularly. WMD had handed over the waste collection to the private companies since 2000 whilst it oversees their operations and currently, it does not provide collection service.

The market share of companies was determined to find out the market concentration. Market share is the ratio of percentage waste collected by each company to the total waste collected in the city. In 2004 and 2005, the company with the highest market share collected the highest waste quantity of 98,369 tonnes and 87,727 tonnes respectively. This represented 24% and 23% of the total waste disposed of at the disposal facility. In 2006, another company became the highest waste collection company disposing of 102,186 tonnes of solid waste, representing 17% of the total waste collected by all the 18 companies. The top three companies had high capacity and more vehicles on road, and these are the reasons why these companies had high quantity of waste collected. Five of the companies had a total of less than 10% and were operating at low capacity with few vehicles on the road (each with less than 3). The analysis of the effect of market share and scale of operation on performance is examined in the subsequent chapter.

Kumasi Coverage and Service Levels

The daily waste generated in Kumasi was about 1200 tonnes, while the average daily waste collected in 2007 was 930 tonnes. Figure 5.9 shows the trend in coverage and private sector involvement and the proportion of waste generation that was actually collected in Kumasi. Collection rate increased from 69% in 2005 to 90% in 2006. The more involvement of private companies has increased the quantity of solid waste collected in the city over the years. Companies also have been gaining experience and acquiring more capacity over time to expand their services. These have influenced the observed increased in the coverage of services. However, if the increase is to be substantial and sustained, KMA efforts towards implementation of pay-as-you-throw (PAYT) to mobilize funds from communal collection system need to be sustained, and also has to ensure good quality of service. The companies also have to be more strategic in their operation plans so as to make the best use of their capacities, and intensify public education campaigns and service promotion.

The communal collection in 2006 was 91% of the waste collected, while house-to-house collection was 8%, and the remaining 1% was disposed of by the industrial sector. The proportions of communal and house-to-house waste collected in 2008 were 89% and 10% respectively. The contribution of the private sector in waste collection is significant, since it collected about 93% of the waste disposed of in 2006. Increasing the capacity and efficiency of these companies would therefore mean improved solid waste collection within the metropolis.

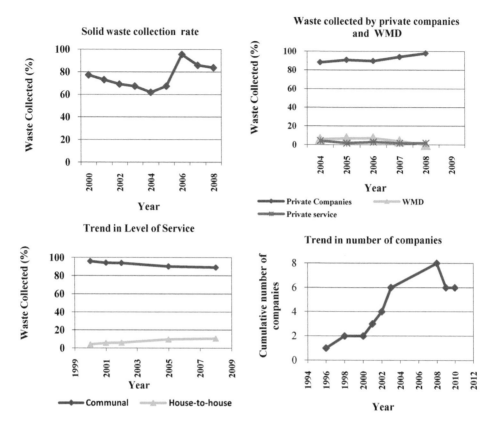

Figure 5.9 Trend in coverage and collection rate and proportion of waste collected by the private companies in Kumasi. Source: Analysis of field data collected by Author

The total waste collected (market share) by the company, which had city monopoly contract in 1999, reduced from 70% in 2004 to 34% in 2006. This company collected 20% of total collected quantity in 2008 as a result of other companies acquiring more capacity to perform. This company is gradually losing its scale of operation in the sector due to increasing number of companies. The reduction in waste collected indicates a corresponding decrease in scale of operation. A company that entered the market in 2004 increased from 6% in 2004 to 36% in 2007 and was the largest collector of solid waste in 2007. A new company that entered the market in 2006 increased from 8% in 2006 to 30% in 2007. This company is now leading with market share of 50% in 2008. The company with the highest market in 2008 has new standard collection vehicles. The decline in the market share of the other companies is due the over-aged vehicles.

Tema Coverage and Service Levels

There was an increase in the waste collection rate (percentage of waste generated that is collected) in Tema from 47% in 2003 to 84% in 2006. The average daily waste collected in 2007 by all companies was 949 tonnes. The high increase in waste collection in the year 2006 could be attributed to the preparation towards the national independence golden jubilee in 2007, because the municipality increased its budget for environmental sanitation in the city. This implies that with enough resources and education or awareness creation on environmental sanitation, solid waste collection in the city can improve drastically.

Figure 5.10 shows the service level, the trend in the cumulative number of companies and the percentages of waste disposed of at disposal site by the private and the public sectors. The public sector collected an average of 14% of the municipal waste, while the private companies collected an average of 63% from 2003 to 2008. The contribution of the private sector in solid waste collection is significant since all service providers involved collected more solid waste (an average 63% of total) in the municipality.

Tema is the only city with high proportion of house-to-house collection (65%) due to the good road network in the city and the income level of the people (mostly middle and high income). The company which had the highest market share (48%) in 2004 is gradually losing its scale of operation in the sector due to increasing number of companies. In 2008, two companies had market shares of 55% in total while three companies, including the largest company in 2004, had market shares of 8% in total, and the Assembly collected 10%.

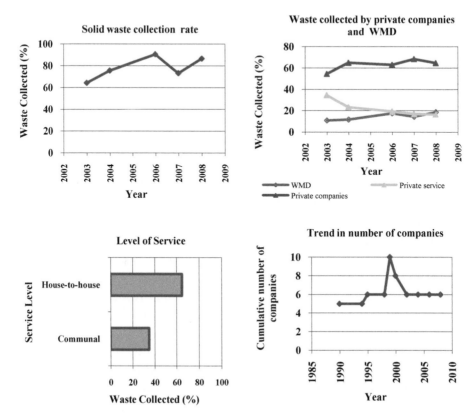

Figure 5.10 Waste collection rates and Disposal by the private companies and TMA in Tema

Source: Analysis of field data collected by Author

Takoradi Coverage and Service Levels

Figure 5.11 shows the trend in waste collection rate and role of private sector in Takoradi. There was an increased in the waste collection rate of regular collection between 2002 and 2005. In 2004, 54.6% of the waste generated in that year was collected, but there was a decrease in 2006 to 45%. There was ground evacuation of accumulated waste on the ground in

2005 and 2006. While population increased over the years and more waste was generated, there was no improvement in the physical capacity and performance of the solid waste companies resulting in accumulation of waste in the city. This is also evident in the quantities of waste collected by the private companies and WMD.

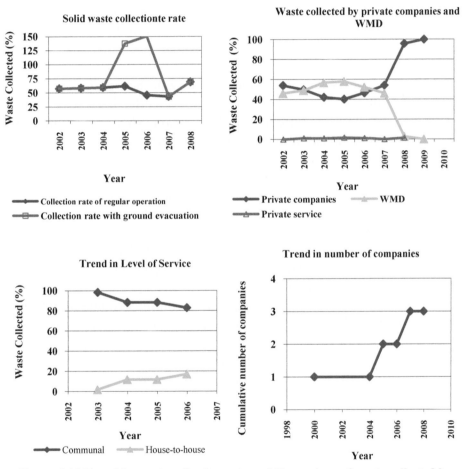

Figure 5.11 Trend in waste collection rate and Percentage of waste collected by house-to-house and communal collection and the private sector in Takoradi
Source: Analysis of field data collected by Author

Within a four year period (2002-2004) the public sector collected 46%, 49%, 56.7% and 58% of the waste collected in those years. The remaining 54%, 50%, 42% and 40% for the respective years was collected by the private sector. These values do not show any significant increase in performance of the private sector. It also indicates that the WMD was collecting more than 20% of the waste specified in the environmental sanitation policy. This can give room for more companies to be introduced into the service to reduce the waste left uncollected which has to be collected by ground evacuation.

Figure 5.11 shows the proportions of house-to-house and communal collection. In 2003, the proportion of solid waste collection was 1.7% for house-to-house and 98.3% for communal collection. There was an increase in the proportion of house-to-house collection from 1.7% in 2003 to 17% in 2006. The slight increase in the house-to-house collection percentage may be as a result of the introduction of tractors with trailer (by the WMD) into the service between 2004 and 2006. The proportion of communal collection decreased from 98% in 2003 to 83% in 2006. From this, it is evident that a greater quantity of the waste generated in the city is collected and disposed of by the communal collection system. This shows the over reliance on communal collection which was free service to the people, but paid by subsidy from general taxes.

The market share of all the companies was on the average of 50% between 2002 and 2007, but increased to about 96% in 2008. This was due to the obsolete collection vehicles and the declining physical capacity of the Waste Management Department. The market share of the first company that entered the market in 2000 declined from 50% in 2005 to 40% in 2008. The market share of the second company increased from 5% in 2006 to 46% in 2008. The market share of the third company is less than 10% for 2008.

Tamale Coverage and Service Levels

Figure 5.12 shows the proportions of house-to-house and communal collection. The house-to-house collection service was carried out by five private companies and represented 12% of collection system in 2006. The communal collection system was done by the Assembly (TAMA) and a private company and represented 88% of total collection.

Before December 2005, there were no records on waste collection by the Assembly. The Assembly was the only sole collector of waste until the introduction of one private company in 2002 under the UESP and five companies in 2005. In 2006, the waste collected and transported to the landfill site was 40% of waste generated. The uncollected waste is usually burnt or buried for composting or dumped at the backyard. It is seen that the Assembly collected 20% of the waste in 2007 while the private companies collected 80%.

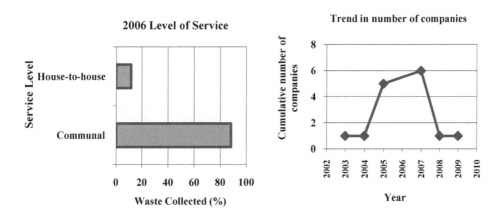

Figure 5.12 Percentage of waste collected by house-to-house and communal collection systems in 2006, and trend in cumulative number of companies
Source: Analysis of field data collected by author

Comparative analysis of trend in service levels, service coverage, and public-private delivery

The main type of collection service in the five cities are the communal collection and house-to-house. Table 5.6 shows the number of companies in the cities and the proportions of solid waste collected in 2007 by the private sector and type of collection. The communal mode of collection was predominant in the cities. The proportions of waste collected by communal collection service in the cities ranged from 35 to 94%. More than 82% of the solid waste collected in four cities was by communal collection mode, apart from Tema (35%). The cost of communal collection is a financial burden to the Assemblies since as of 2008 the customers were not paying in Accra and Kumasi, and as of now they are not paying for the service in Takoradi and Tamale. There is an initiative in Kumasi, Accra, Tema, and Takoradi toward PAYT and flat fee, but there is the need to improve the process by involving the households to increase the level of awareness and make the private sector more responsive to the customers to realise the policy on full cost recovery.

Table 5.6 Percentage of waste collected by public and private sectors and collection mode in 2007

City	Number of formal private companies	Solid Waste Collection by		Level of Service	
		Public sector (%)	Private sector (%)	House-to-house (%)	Communal (%)
Accra	18	0	100	6	94
Kumasi	8	5	95	18	82
Tema	7	22	78	65	35
Takoradi	3	45	55	17	83
Tamale	6	26	74	12	88

Source: from analysis of field data collected by author

The house-to-house solid waste collection provided was less than 20% of waste collected in four cities, but high (65%) for Tema due to its planned settlement nature and income level of the people. House-to-house collection is a method of domestic solid waste collection in which the individuals place bins full of waste outside their houses at the curb side or roadside on specific days for collection. The residents served by the house-to-house waste collection use standard bins (80, 120, 240 litres). The house-to-house service is mostly rendered in the high and middle income areas.

Table 5.7 shows the solid waste collection rate. The collection rate is the proportion of solid waste generated which is actually collected and disposed of from the city. The solid waste collected used in the analysis excludes the waste material that is generated but does not enter the municipal waste stream because it is reused by residents or collected from residents and sold separately by itinerant waste buyers and collectors. The introduction of the private sector in solid waste collection increased the proportion of waste generated which was collected (collection rate) from range of 10%-60% in 1997 and to the range of 57% to 77% in 2004. The collection rate increased as a result of the introduction of PSI by World Bank's Urban Environmental Sanitation Project (UESP). There has been an increased PSI between 2004 and 2008 with more companies entering into solid waste business. However, the collection rates of two cities (Tamale and Sekondi-Takoradi) decreased between 2004 and 2007, despite high PSI in Tamale. The capacities of the companies in Tamale were low and did not match the quantity of waste generated. This suggests that high levels of PSI couple with low capacity (as in the case of Tamale) in terms of collection trucks would not lead to increase collection rate or coverage.

Table 5.7 Waste collection as a percentage of waste generated

City	2000 Population	1997 (%)	2004[1] UESP (%)	2006[2] (%)
Accra	1,700,000	60	63	80
Kumasi	1,200,000	45	77	91
Tema	506,000	50	60	84
Takoradi	400,000	40	72	55
Tamale	275,000	10	57	40

[1]*Outcome of UESP-Urban Environmental Sanitation Project (1997-2003) from Boakye (2005); [2]Analysis of field data collected in 2007 by the author.*

There is an increasing role of the private sector in solid waste collection service. More than 60% of the solid waste in the cities was collected by the formal private sector. According to the Ghana Environmental Sanitation Policy (1999), the Waste Management Department of the Assembly is mandated to maintain an in-house capacity to provide twenty percent (20%) of the solid waste services directly and outsourced the 80% to the private sector. The public sector in Tamale and Sekondi-Takoradi collected 26% and 45% of solid waste (Table 5.6) respectively as against at most 20% stated in the policy. The degree of private sector involvement depends on the in-house capacity and resources of the Assembly to provide direct service. The Assembly with limited in-house capacity requires high level of PSI with adequate capacity to increase collection rate. For example, among the five cities, Sekondi-Takoradi had the lowest PSI and low collection rate (55%). The city with few companies (Sekondi-Takoradi) had 55% PSI whereas the largest city (Accra) with 18 companies had 100% PSI.

The Assemblies and the companies are however faced with the problems of inadequate waste collection equipment and containers, lack of funds to procure new trucks, lack of political will to support cost recovery of collection service and unwillingness of some residents to contribute towards cost recovery of waste collection. The low cost recovery has led to financial constraint and therefore Metropolitan Assemblies are unable to keep and maintain the cities clean as expected.

The implementation of the policies on fair competition, engaging companies with signed contracts, and packing all services into one contract has been slow. The environmental sanitation policy (MLGRD, 2008) stipulates the provision of communal service as part of a franchise contract, fair competition among private sector service providers and the engagement of service providers in accordance with the laws of Ghana and the Public Procurement Act. The Assemblies are now engaging service providers in accordance with the laws of Ghana and the Public Procurement Act (2003) which demand competitive bidding and signing of contract.

5.3   Households' Involvement in Solid Waste Collection

This section presents the households' involvement in service provision in terms of their contribution towards cost recovery, willingness-to-pay for service, willingness to separate waste at source, and service quality monitoring.

5.3.1   Cost Recovery and Willingness-to-pay

Cost recovery Mechanisms

There was no involvement of the private sector in the setting of tariffs by the Assemblies. The tariffs (service charges and collection fees) set were low, and upward review always delay over 2 years. The policy (MLGRD, 2008) stipulates that the private sector investment shall be encouraged by a flexible approach to public-private partnership arrangements and service provision subject to the supervision and setting of maximum tariffs by the Assemblies with full participation by private sector and users. The policy states that direct cost recovery from all users should be applied where it is possible to charge a full commercial price covering all operating and capital costs of solid waste collection but where full direct cost recovery is not possible, the shortfall shall be subsidised by the Assembly. It is therefore essential that levels of service are demand driven and mechanisms for funding them are sustainable. The recent shift toward cost recovery is an effort towards the implementation of the policy on cost recovery after ten years of existence of the policy. The pay-as-you-throw (PAYT) mechanism is appropriate for low income communities compared to fixed monthly fee.

The involvement of households in paying for solid waste collection to reduce the financial burden on the Assemblies started with the introduction of franchise house-to-house collection. Prior to 2009, most users of communal collection did not pay for dumping into communal containers, and the involvement of communal service recipients to pay for service has been slow until recently. Apart from Tema where communal service recipients pay small monthly fee, less than 20% of the people in the other cities pay for house-to-house service, but most people (80%) who receive communal service did not pay for service as of 2008. There are modalities for the partial and full cost recovery through charging monthly flat charge from end users of house-to-house service, but the implementation has always been a problem. The pay-as-you-throw (PAYT) basis for communal service was introduced in 1995 in Accra but it failed. The PAYT initiative in Kumasi could be a model to learn from and replicate in the other cities.

A shift towards cost recovery to charge end users on pay-as-you-throw (PAYT) basis for communal service and monthly fixed charges for house-to-house (or kerbside collection) began recently with the city-wide-fee for solid waste collection in Accra, Kumasi and Tema. There is an initiative from 2008 in Kumasi, Accra, Tema, and Takoradi to involve communal service recipients to pay either flat fee every month or pay-as-they-throw (PAYT) waste away at communal container site. Charging users for the service is the new approach (city-wide-fee) for cost recovery and service delivery in Kumasi and Accra. The city-wide-fee (PAYT) approach in Kumasi initially covered 50% of the 150 container sites. A subsidy of GH¢ 20.00 (US$ 15.00) per container of 23m^3 (40% of cost) is paid to companies for every trip and an amount of GH¢30.00 (US$ 23.00) per container of 23m^3 (60% of cost) is collected from user charges through the PAYT. This has reduced the financial burden on the Assembly by 60% which hitherto was borne by the Assembly. The survey conducted in Kumasi revealed that the people in respective localities were not consulted for their views before pay-as-you-throw (PAYT) was introduced. It was evident from the survey that 80% of people interviewed had no idea about PAYT prior to the implementation of the PAYT. The community involvement in PAYT before introduction was very poor. Community forum and meetings which provide platform for expression of views during development processes to give the people 'choice' and 'voice' were not done.

There are problems with the implementation of Accra city-wide-fee for solid waste collection. The approach is that all households including low income groups are required to pay monthly fixed service fees to be collected by the companies. The service fee is intended to cover both collection and disposal cost without subsidy from Assembly. However, there are some problems with this system especially with the poor communities. This study shows that PAYT where residents pay 10, 20 or 30 Ghana Pesewas (US$0.06-0.20) depending on the quantity of waste to be dumped could be a better option of user charging for low income communities, instead of fixed fee per household which is collected monthly. PAYT does not accumulate the charges compared to the monthly fixed fee being pay by poor urban households. The results from the Kumasi study show that poor households could pay as low as GH¢1.50 (cumulative PAYT of GH¢0.10 for a month) instead of monthly fixed fee of GH¢3.30. The problems with the city-wide-fee for solid waste collection in Accra are: (1) financial burden on the companies to distribute over 10,000 bins to all households (2) unwillingness of low income households to register for collection, (3) there is low payment rate at the low income communities, (4) illegal and indiscriminate dumping by some households who refused to register, (5) there is also high disposal fee (tipping fee) to be paid from the service charges collected from the households, and (6) weak enforcement of non-registration and illegal dumping by the Assembly.

The involvement of communal collection service recipients to pay for the service through pay-as-you-throw (PAYT) was initiated by some Assemblymen in Accra but the collection of PAYT user charges at container points was unofficial. The collection of the PAYT charges in Accra prior to the city-wide-fee collection was illegitimate, since the collection of user charges was not approved by the Assembly's fee fixing resolution. The Assembly has a constitutional right to fix and collect user charges. There was involvement of the Assemblymen (and Assemblywomen) in the collection of illegal user charges. This illegality was confirmed in key informant interview with officials from the WMD. The survey conducted in Accra reveals that the people receiving communal collection service were not consulted for their views before PAYT was introduced by the Assemblymen. All the residents interviewed had no idea about PAYT and some were aware that it was illegal. There was a similar situation in Takoradi, where some Assemblymen were involved in the collection of PAYT charges from those who use the communal containers.

Households' Perception on Cost Recovery Options

Table 5.8 presents the households' perception on options for cost recovery (opinion on who should pay for the waste collection) across the five cities. Most of the respondents were aware of their role to pay towards the solid waste collection to improve the service quality. The survey showed that 70% of the respondents across cities who received house-to-house collection service were of the view that the generators, Assemblies and the Government should contribute towards recovery of the cost incurred in waste collection and disposal, whereas 15% said the Assemblies should pay for waste collection service.

Table 5.8 Opinion on who should pay for the waste collection services

Cost recovery options	N	Accra	Kumasi	Tema	Ta-male	Ta-koradi	%
House-to-house collection		N=362	N=133	N=157	N=61	N=125	
Generators only	107	58	0	2	7	40	13
Generators and Assembly/government	584	278	128	83	42	53	70
Government and Assembly	129	19	5	71	4	30	15
No response/cannot tell	18	7	0	1	8	2	2
Communal collection		N=166	N=138	N=27	N=19	N=93	
Generators only	14	1	1	0	0	12	4
Generators and Assembly/Government	216	93	48	10	11	54	55
Government and Assembly	153	19	89	16	8	21	39
No Response/Cannot tell	10	3	0	1	0	6	3

Results of field survey by Author in 2007

For the communal collection, 55% of the respondents across cities were of the view that cost should be collectively paid by generators, Assemblies and the Government, whereas 39% said the Assemblies should pay for the collection cost. Those who think only generators should pay

were few, 13% of house-to-house respondents and 4% of those served under communal collection. This presupposes that more effort has to be put
into household involvement and the companies made to be accountable for
results and be more responsive to the user. This will make the households
aware of the financial burden on the Assemblies and the need to pay for
service improvement.

The Assemblies over the years have offered the communal collection
service free of charge, and therefore faced financial constraints which affected the level of efficiency of waste collection as well as service quality.
There is the need to improve the process by involving the households to pay
for the service and to increase the level of awareness, and make the private
sector more responsive to the costumers to realise the policy on full cost recovery.

Willingness-to-pay

Besides the analysis of households' perception on options for cost recovery, their willingness to pay more was assessed. Table 5.9 shows the responses of the respondents' willingness-to-pay user charges. The willingness-to-pay more to improve house-to-house service was low as a result of the
service recipient perceived dissatisfaction with the current service quality.
The results of willingness-to-pay more for house-to-house service show that
20% of respondents were willing to pay current charges for the service,
whiles majority (62%) were willing to pay less than current charges. They
perceived the service quality to be poor and that they would like to see improvement in service delivery first before they will pay more. Besides,
some respondents (18%) were willing to pay more provided there will be
improvement in service and increase in the quality of service.

The respondents willing to pay more or same or less were demanding
value for money. They were willing to pay appreciable money for the service if the service is improved in terms of regular collection and twice a
week collection frequency. They said the user charges should be based on
collection frequency (twice collection), waste volume, and for regular collection instead of the monthly fixed fees. Those willing to pay more were

172 | Chapter Five

of the view that there is still room for service improvement. The willingness to pay more was high in the high income areas compared to the middle income areas. This means that customer orientation and accountability
for results and better service quality are necessary for increasing service
charges. The private sector needs to be responsive to the customers and to
provide better service quality, and this need is affirmed by some authors
(World Bank, 1997; Gidman et al., 1999).

Table 5.9 Respondents' willingness-to-pay user charges

Willingness to pay by house-holds	N	Accra	Ku-masi	Tema	Ta-male	Ta-koradi	%
House-to-house collection	819	N=354	N=133	N=157	N=57	N=118	
Does not pay user charge	45	0	0	3	41	1	5
Pays user charge	774	354	133	154	16	117	95
Willingness to pay more	144	89	31	0	24	0	18
Willingness to pay current user charge	165	95	17	17	6	30	20
Willing to pay less than current user charge	510	170	85	140	27	88	62
Communal collection	380	N=112	N=138	N=22	N=19	N=89	
Did not pay user charge at time of survey	202	9	135	15	16	27	53
Pays user charge at time of survey	178	103	3	7	3	62	47
Willingness to pay for the service	161	5	126	2	12	16	42
Willingness to pay less or not willing to pay	219	107	12	20	7	73	58

Results of field survey by Author in 2007

The results of willingness-to-pay for communal collection service show
that 53% of the respondents across the cities at the time of survey did not
pay and 42% of respondents were willing to pay for the service. There was
high degree of willingness-to-pay for the collection service in Kumasi
(91%) and Tamale (63%) but low in Accra (5%) and Takoradi (18%), even
though most of the respondents in Accra and Takoradi were paying unofficially for the waste they throw away. The survey shows that 47% of the
respondents were paying-as-they-throw waste into communal containers.

This was an illegal collection of user charges in Accra (92%) and Takoradi (70%). The Assemblies were aware of the illegal collection of user charges by some Assemblymen and residents but nothing was being done about it. Even though the PAYT in Accra and Takoradi was not accounted for to the Assemblies, the people were willing to pay and therefore showed the level of cooperation from the people. They were willing to pay as and when they go to dump their waste instead of monthly payment, but according to Obirih-Opareh and Post (2002) pay-as-you-throw (PAYT) did not work in Accra in 1995 as it encouraged indiscriminate dumping to avoid payment by children who send the waste.

The gradual growth of house-to-house (HtH) collection services directly implies gradual growth of cost recovery. It is in the interest of the Assemblies to support the growth of the HtH service in all middle income communities since increasing the HtH coverage directly relates to a decrease in the communal collection. The sustainable SWM can only be realized if cost recovery policy is pursued. The charging of user fees from CCC users to increase cost recovery through pay-as-you-throw to overcome the financial burden of SWC is necessary.

5.3.2 Households' Awareness of Waste Minimization

Source Separation, Waste Reuse and Recycling

Table 5.10 presents the willingness-to-source separate organic and recyclable waste at the house for separate collection. The results show that 48% of the respondents across the cities were willing to source separate waste. The respondents who were willing to separate organic waste for separate collection in Kumasi, Tamale and Accra were 91%, 68% and 55% respectively. Those who were willing to separate indicated a number of incentives needed for separating waste for collection. These include provision of free polythene bags of different colours, increased collection frequency to twice a week, some agree to buy their own bins for storing organic waste and pay reduced fee, and some want free bins to be provided by companies and assembly for separate collection. Some suggested that there must be massive educational campaign and that the separation be made by-law to compel everybody to separate. Others were not willing to separate waste because of the perception about waste in the communities, that it is time wasting to separate. Most of the respondents in Tema and Takoradi were undecided. The education and motivation of households could play a role in creating the necessary awareness about source separation of waste.

With regard to waste minimisation, most respondents were aware of waste reuse and recycling, and some knew the benefits of composting. Most of the respondents said they reuse empty plastics and cans, and sell plastics containers and sachet bags/plastics to buyers and scavengers. They also engage in the burning of combustibles. Few of them use organic waste for making compost for backyard garden and manure. Despite the reuse and recycle of these waste fractions, greater amount of the reuse and recyclable materials are found in the waste stream which eventually are contaminated and find their way to the final disposal site. The system of households' solid waste collection does not encourage recycling. Mixed waste is collected and sent to disposal sites and some form of recycling goes on at the dumping sites.

Table 5.10 Number of respondents willing to separate organic waste at the house for separate collection

	House-to-house collection			Communal collection				Combined (%)	
	no	yes	undecided	no	yes	undecided	N	No	Yes
Accra	99	201	62	27	57	32	478	28	55
Kumasi	3	130	0	6	132	0	271	3	91
Tema	0	45	112	0	0	27	184	0	24
Tamale	1	43	17	2	11	6	80	4	68
Takoradi	7	4	114	4	5	84	218	5	4

Results of field survey by Author in 2007

There is awareness of waste reuse and recycling, what is needed is the formal arrangements for source separation and the collection of these recyclables. Most of the companies have no plans to undertake source separation and waste recycling since there are no economic and legal incentives to do so at the moment. With the initiative of city-wide-fee for solid waste collection, there will be the need to enact legislative instrument to regulate recycling and reduce organic fraction of waste sent to landfill as well as allowing private sector to own and manage landfill. This will provide the necessary incentives for private sector involvement in waste recycling and recovery to ensure sustainable development. Only one company has started the construction of compost and sorting plant to handle waste from Accra. The compost from the plant is intended to be sold to companies involved in landscaping. The management of another company has intention to construct a compost plant in Accra.

Awareness of waste impact and prevention of indiscriminate dumping

Table 5.11 presents the households' awareness of prevention of waste being dumped into drains and other unauthorised places. Results show that some households opted out of the formal collection service and most respondents are aware of measures to deal with illegal dumping of waste into drains and other unauthorised places. The study reveals that there was no complete collection of waste from all households in communities where house-to-house is served, since some of the households (16%) opted out and were not served in areas where companies operate, especially in the middle income areas. The proportion of the 1434 respondents who practice open dumping (self disposal) across the cities was 11%. In communal collection communities some of the people do not send waste to the communal containers but rather dump it anywhere. These people were not involved in paying for the service because the by-law enforcement by the Assembly was weak.

Most respondents are aware of measures to deal with illegal and indiscriminate dumping of solid waste. These measures are categorised into three: collection service improvement, enforcement of by-laws, and public education. The households are aware of the environmental problems but they need to be involved in the collection to bring about attitudinal and be-

havioural change. This can be done through public education as some suggested and accountability of the companies for results. Indiscriminate
dumping is common in the middle and low income areas where communal
collection service is rendered. The communal skip containers are accessible by residents and sites were not too far from the places of residence of
those who use them. The housing types and family structure are such that
many people live in a house which is usually two to three storey or compound houses, and the layouts of the areas do not allow house-to-house
collection of waste and as such people resort to sending their waste to a
communal site.

Table 5.11 Awareness of prevention of indiscriminate dumping of waste

Awareness of prevention of indiscriminate dumping	Accra	Kumasi	Tema	Tamale	Takoradi
House-to-house collection	(N=362)	(N=133)	(N=157)	(N=61)	(N=125)
Collect waste every day, provide more bins and supervise	17	50	52	18	9
Bins should be emptied at the right time	15	7	67	4	18
Enforce by-laws and defaulters should be sanctioned	182	7	1	10	33
Educate the public	39	24	6	19	1
Drains should be covered and improve service	36	37	26	5	24
No response/cannot tell	73	8	5	5	40
Communal collection	(N=116)	(N=138)	(N=27)	(N=19)	(N=93)
Collect waste every day, provide more bins and supervise	9	60	9	3	31
Containers/bins should be emptied at the right time	1	2	4	0	6
Enforce by-laws and defaulters should be sanctioned	72	10	0	5	9
Educate the public	10	38	3	5	4
Drains should be covered and improve service	21	27	9	6	13
No response/cannot tell	5	1	2	0	30
Self disposal of waste N	482	290	225	152	285
(%)	0.6	7	17	45	10

Results of field survey by Author in 2007

The results show that more than half the respondents were aware of the impact of uncollected waste on neighbourhood aesthetics and human health. They said waste at some communal sites gives offensive odour in the neighbourhoods. Where there were no standard bins, waste placed in uncovered container is littered all over the place so when it rains, the waste is washed into nearby drains, carried away and affects the aesthetics of the city. Residents also deliberately dump waste into drains during rainfall. This chokes the drains and results in breeding of mosquitoes and flooding in certain parts of the cities after subsequent rains. The indiscriminate dumping of waste results in serious health implications, and the most prevalent is malaria as indicated by most of the respondents. This has a direct effect on the productivity of the workers who acquire the disease. As a result stringent measures on by-law enforcement would be needed by the city authorities to curb the social menace of indiscriminate waste disposal into drains and open spaces.

The littering at the sanitary sites is due to the lax attitude and lack of co-operation of the people, and irregular waste collection. When people go to dispose of their waste and find out that others have dumped theirs on the ground, they do likewise because they do not see why they should walk through the refuse to dump theirs in the container. The respondents attributed the problems of overflowing of waste from skip containers onto the ground at the communal sanitary sites to irregular collection of containers from sanitary sites, no watchmen at night or lack of attendants at container sites, poor positioning of skip containers by truck drivers, inadequate support for children who send waste into containers, and negative attitude of inhabitants. Some containers were not collected on time when they were full, thus contributing to the general insanitary conditions of some the suburbs in the cities. The problems of environmental pollution could be attributed to lack of public awareness of the health implications of insanitary practices, residents' indifference to the presence of waste and unclean environment, inadequate trucks for frequent collection of containers. All the Assemblies did not collect all waste leaving about 10% to 30% of the waste generated which is not collected making the environment aesthetically unpleasing. Those greatly affected are people who live in low income areas who also do not pay for collection service and dump indiscriminately.

The implementation of the policy on regulation of waste recycle and enforcement of by-laws is weak due to the low capacity of the Assemblies and the political will to implement the policies. The policy (MLGRD, 2008) stipulates re-structuring of institutional arrangements and legal framework, strengthening of the capacity of law enforcement agencies to deal with environmental sanitation. The need for legislation, enforcement and regulation for pollution prevention, waste recycling and treatment has not received the needed attention. The policy required the Assemblies to promulgate by-laws consistent with the model by-laws. The policy also mentions the need for providing legislation and incentives for waste reduction, re-use, recycling and recovery; as well as enacting, disseminating and enforcing legislations/by-laws that prohibit illegal dumping of waste and govern sector-wide standards. This study clearly shows that despite the existence of the policies for a decade now not much has been done to enforce by-laws on illegal dumping and to regulate waste recycling through legislative and incentive provisions. With the gradual shift towards full scale cost recovery and the availability of environmental sanitation action plan, it is necessary that the enforcement of the by-laws and regulation of waste recycling and recovery be taken on board seriously.

Households' involvement in service quality monitoring

Table 5.12 presents the respondents' awareness of waste container overflows before container is lifted. More than half of the respondents served under communal service (85%) and house-to-house service (54%) said waste overflows before collection. The level of satisfaction about the collection service provided by the private companies was rated based on the frequency of collection and waste overflow. There was dissatisfaction about the collection service for communal users due to the fact that the frequency of waste collection was very low and as such the waste containers were always overflowing. All the respondents who were not happy about the collection service said that the haulage trucks for communal and the house-to-house collection were not regular at all, and as a result the skips for communal collection and the standard bins for house-to-house were all the time overflowing, thus causing a lot of nuisance and insanitary conditions. Uncollected heaps of refuse are aesthetically unpleasant, and create breeding places for mosquitoes and flies. The respondents admitted the sanitary conditions of the surroundings of communal containers as poor due to the uncollected heaps of refuse, blocked drains and the negative health impacts.

Table 5.12 Awareness of waste container overflows before container is lifted

	House-to-house collection			Communal collection		
City	N	Yes	No	N	Yes	No
Accra	312	216	96	105	101	4
Kumasi	132	60	72	136	120	16
Tema	145	114	31	23	14	9
Tamale	51	9	42	17	2	15
Takoradi	103	5	98	76	68	8
Total		404	339		305	52

Results of field survey by Author in 2007

More than half of the respondents of the house-to-house service were not happy about the quality of service, but there is no formal channel for service quality monitoring through customer complaints and responses to complaints as required by market orientation. There is waste spillover from bins and this makes their houses aesthetically unpleasant. They have lots of complaints but there is no channel (office and phone numbers) for such reports. The respondents were of the view that the service providers do not have simple tools such as brooms, rakes or shovels and as such they leave spillover waste in front of their houses which makes the place untidy when they come to collect refuse. They were of the view that additional bins should be provided since sometimes the refuse is not collected as scheduled. The scheduled time for collection is twice a week although the refuse is collected once a week. The collection frequency of once a week was common for house-to-house and daily and every 3 days collection for communal collection but more than half mentioned that waste was uncollected for a week or more, and the collection frequency was not consistent.

The respondents perceived accumulations of solid waste at the communal site to be a problem. The skip containers were not collected on time so the waste brought by inhabitants was dumped on the ground of the container site when the container is full. The irregular collection of containers was due to inadequacy of trucks and at times breakdown of vehicles.

5.4   Conclusion

This chapter explored and discussed the evolving practices of involving private sector companies and households in service provision, and the policy implications. This chapter has shown how private sector has evolved over the past two decades and the extent of implementation of environmental sanitation policy. The analysis shows that there is an increasing role of the private sector in solid waste collection service over the past two decade with each company operating in a defined collection zone. More than 60% of the solid waste in the cities was collected by the formal private sector. The national policies for solid waste collection in Ghana have changed from local government provision to private sector provision, but still more has to be done to improve service delivery. The private sector involvement has evolved from company entry into waste collection business with or without subsidy and contract through to arrangements between public sector and private companies with contracts and cost recovery arrangements.

Notwithstanding the zoning of cities into solid waste collection zones and increasing involvement of the private sector, the involvement of households to pay for service provision has been slow (76% did not pay for service in 2006) and these are those who receive communal service. Results show that 20% of respondents served with house-to-house were willing to pay current charges for the service, while 62% were willing to pay less than what they are paying. Results show that 53% of the respondents served under communal collection service at the time of survey did not pay anything officially. Forty-two percent (42%) of communal service respondents across cities were willing to pay. The proportion of households who pay for waste was low and this places financial burden on the municipalities. The orientation of customers, accountability for results and better service quality are necessary for increasing service charges. This finding confirms the hypothesis that increased private involvement without corresponding increase households' involvement do not ensure that the private sector becomes responsive to the customers and provides better service quality.

The involvement of communal collection service recipients to pay for the service was initiated by some Assemblymen but the collection of user charges at container points was not accounted for. There was an involvement of the Assemblymen (and Assemblywomen) in the collection of illegal user charges in Accra. The Assemblymen in Takoradi were tasked in 2007 to collect user charges and render accounts to the Assembly. Unfortunately, most of the Assemblymen did not implement the PAYT because of the fear of losing political re-election into the Assembly and those who collected the charges did not account for the revenue.

A shift towards cost recovery to charge end users on pay-as-you-throw (PAYT) basis for communal service and monthly flat charges for kerbside collection began recently with the city-wide-fee for solid waste collection in Accra, Kumasi and Tema. There is an initiative from 2008 in Kumasi, Accra, Tema, and Takoradi to involve communal service recipients to pay either fixed fee every month or pay-as-they-throw (PAYT) user charges at communal container site. The analysis show that more than 60% of the solid waste collection cost is paid by PAYT revenue in Kumasi. Therefore, if properly managed and the cost accounting properly done, PAYT could be a source of revenue for full cost recovery to replace subsidies being paid to the private companies. The PAYT approach was used in Kumasi for the low income communities, whereas in Accra monthly service fee was imposed on low income households. The problems companies in Accra encountered were financial burden to distribute bins, unwillingness of households to register for collection, low payment rate at the low income communities, illegal and indiscriminate dumping by some households who refused to register, high disposal fee (tipping fee), and weak enforcement of registration of non-registered households by the Assembly.

With regard to service quality monitoring and waste minimisation, households are aware of inefficient services, cost recovery options, waste reuse and recycling, what is needed is the formal organised arrangement for source separation and collection of these recyclables. Most of the companies have no plans to undertake source separation and waste recycling since there are no economic and legal incentives to do so. More than half of the respondents of the house-to-house service were not happy about the quality of service, but there is no formal channel for service quality monitoring through customer complaints and responses to complaints, as required by market orientation.

The next chapter analyses and examines the productivity and service quality of private companies involved in solid waste collection and the effect of the scale of operation or company size on performance.

6 Private Sector Performance: Productivity and Service Quality in Solid Waste Collection

6.1 Introduction

Chapter 5 examined the evolving forms of private sector involvement in urban solid waste collection. It was identified that the companies across the cities have varied market share of waste collected and scale of operation. This chapter analyses the performance of the private sector companies in terms of productivity and service quality. The differences in service operations between two groups of companies (those below and above industry average performance) will be explored. These differences will bring to the forefront the issue of capacity of the companies. This study seeks to explore the influence of capacity (scale of operation and capabilities) and regulation on the performance of the private sector. The relationships between productivity and service quality on the one hand, and capacity and regulatory mechanisms on the other hand will be explored in chapters 7 and 8.

This chapter is organised into six sections. Section two provides the background of the productivity and efficiency analyses. Section three presents the results of the productivity of the companies using data envelopment analysis (DEA) and ratio analysis. Section four presents the comparative analysis of productivity across cities.e Section five analyses the households' appreciation of quality of service of the companies. The last section draws summary and conclusion from the analyses and discussions on performance (productivity and service quality) of the companies.

6.2   Productivity and Efficiency Analysis

Data Envelopment Analysis

Productivity is an efficiency measure of performance of companies. The efficiency measurement at firm level using data envelopment analysis (DEA) is one approach used by many scholars to measure firm's productivity. The data envelopment analysis methodology compares firms, departments or organisations. DEA compares the efficiencies in these organisations which are called Decision Making Units (DMUs). DEA compares each company with other companies and identifies those companies that are operating inefficiently compared to best practice or relative efficient companies called efficiency reference set. DEA is an operational research tool that measures efficiency indices by solving linear programming problems – which seek to maximise output of service (or products), or minimise input of resources. The basic concept behind DEA is that an output frontier of the most efficient DMUs is defined and is then used to measure how far are the less efficient DMUs from the frontier (Cooper et al., 2000). DEA determines technical efficiency which is a measure of both scale inefficiency and pure technical inefficiency. Two sources of efficiencies in company operations are measured by the pure technical efficiency and the scale efficiency separately.

Technical Efficiency = Pure Technical Efficiency × Scale Efficiency

The technical efficiency measure compares the actual output of a company to that of another company at a point on the output frontier that uses the same amount of input. The most technical efficient company will have a technical efficiency value of one (100%) and the inefficient ones have technical efficiency less than one. The pure technical inefficiency arises from inefficient operations of the DMU itself (Cooper et al., 2000). The efficient operations of a company depend on some factors such as skills and capabilities of workers and conditions of the solid waste collection trucks. The scale inefficiency arises from sub-optimal scale of operation under which the DMU is operating (Cooper et al., 2000). The scale of operation in this study is taken to be the market share and number of collection vehicles. In this study, the efficiency measures were obtained for each company by conducting DEA using output-oriented variable returns to scale. The output-oriented model maximises the output for given input. The output and input used in this analysis are the quantities of solid waste collected and number of vehicles of each company. The DEA provides efficiency measure for comparing all companies.

Ratio Analysis of Productivity

The performance in terms of productivity and vehicle utilisation of a company was assessed using ratio analysis and the results compared with those of the DEA. This was done to provide an explanation of the variations in productivity. The productivity value from ratio analysis was computed for each company for the purpose of comparison with DEA results and interpretation of the variations in performance across companies. The productivity of each company was computed using weighted average of the productivity of the individual vehicles. The productivity is defined as the ratio of output (waste collected in tonnes) to input (vehicle). Another measure of efficiency is vehicle utilisation, which is defined as the ratio of actual waste collected (output) to company capacity or maximum output of the input. The following mathematical expressions were used to compute productivity and utilisation.

The productivity of an individual vehicle *v* is defined as the average quantity of waste output *q* for a period *t* as follows:

$$P_v^t = \frac{\sum_{q=1}^{Q} W_q^t}{\sum_{t=1}^{T} t}$$

Where:

P_v^t = productivity of vehicle v during the period t, t=1,...T;

W_q^t = actual quantity of output *q* for vehicle *v* during the period t, t=1,...T; q=1,.....Q

t = the working time in days for vehicle *v* during the period t, t=1,...T

The productivity of a company *c* is defined as the weighted average of individual vehicles *v* for a period *t* as follows:

$$P_c^t = \frac{\sum_{v=1}^{V} P_v^t e_v^t}{\sum_{v=1}^{V} e_v^t}$$

Where:

P_c^t = productivity of company *c* during the period t, t=1,...T; v=1...V

e_v^t = maximum output capacity for vehicle *v* during the period t, t=1,...T; v=1...V

Vehicle utilisation is defined as the ratio of actual waste collected to company capacity or maximum output. The capacity utilisation is defined as follows:

$$U_c^t = \frac{P_c^t}{\sum_{v=1}^{V} e_v^t}$$

Where:

U_c^t = capacity utilisation of company c during the period t, t=1,...T;

In this study, the performance of a company was assessed in terms of productivity, vehicle utilisation, and technical efficiencies derived from data envelopment analysis (DEA). Data for the determination of these efficiency measures were obtained from the Municipal Authorities in the five cities. Daily records on number of trips, tonnage of each trip, quantities of waste collected and vehicle types were obtained for each of the 33 companies for a period of three months.

6.3   Variations in Productivity and Efficiency

Results of Data Envelopment Analysis

Figure 6.1 shows the technical efficiencies of companies in the cities. The DEA results identified the relatively efficient best practice company (C15), the less productive companies and the magnitude of inefficiency (use of excess resources) of less productive companies compared to the best practice companies. The results showed that only one company (C15 in Tema) had technical efficiency rating of 100%. Fourteen (14) companies had efficiencies above 50%, and the rest had technical efficiencies less than 50%. In DEA, any shortfall of a company from the 100% technical efficiency mark means there is inefficiency or use of excess resources equivalent to the shortfall in efficiency. For example, company C2 in Accra was less productive with technical efficiency rating of 17%, suggesting that it could collect the current output (quantity of solid waste) with only 17% of the resources it actually used. Thus, C2 was 83% inefficient or was using 83% excess resources. Company C4 in Accra had an efficiency rating of 67% indicating that it was using 33% excess resources.

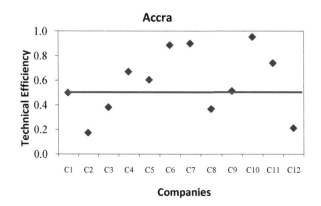

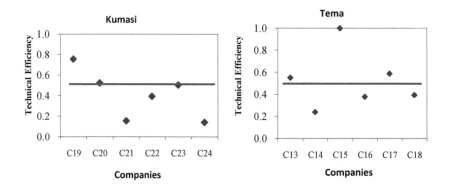

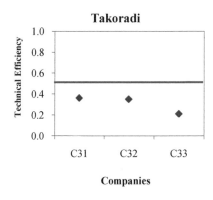
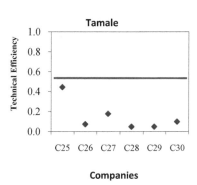

Figure 6.1 DEA Technical efficiency of companies in the cities, showing inefficiencies arising from inefficient operations and scale inefficiencies

Table 6.1 presents the summary of technical efficiency results for the 33 solid waste companies. The technical efficiency results showed that 19 of the companies were less productive with efficiency rating of less than 50%. Thus, 14 of the companies were above an average of 50%. All the 19 companies were using more than 50% excess resources. A total of 25 companies were using 40% or more excess resources, and 7 companies were using 30% or less excess resources (Table 6.1). All the 9 companies in Tamale and Takoradi were all below the average and had more than 50% inefficiencies. The reasons for the inefficiencies or using excess resources could be attributed to inefficient operations and sub-optimal scale of operation. The inefficiencies due to company capacity will be further explored in the next chapter.

Twenty-nine (29) of the companies were scale inefficient compared to their reference sets. The inefficiencies of these companies may be due to pure technical inefficiencies and/or scale inefficiencies. Companies C10 and C18 were reference set for 19 companies. Four companies C7, C10, C15 and C18 had no other companies as reference set. They are their own reference set and had pure technical efficiency of 100%. This means that the operations of these companies were efficient but the scale of operation may not be optimal since their scale efficiencies were less than 100%. Company C2 was inefficient compared with C18, its efficiency reference set. Company C11 with technical efficiency rating of 74% means it was more efficient than C12 with a technical efficiency rating of 21% because they both had the same reference set.

Table 6.1 Number of Companies by DEA Technical Efficiency

Cities	N	With 100% efficiency	With 80-99% efficiency	With 60-79% efficiency	With 50-59% efficiency	With less than 50% efficiency
Accra	12		3	3	2	4
Kumasi	6			1	2	3
Tema	6	1			2	3
Takoradi	3					3
Tamale	6					6
All cities	33	1	3	4	6	19

Companies which are technically inefficient could be able to increase output while using the same inputs. An efficiency rating of 74% for C11 suggests 26% inefficiency in its operation, which tends to be more than the inefficiency of 10% noted in C7 with efficiency of 90%. Both companies have potential productivity improvements but the actual improvements that will ultimately be achieved may be greater in company C11 than C7. The companies could collect the current output quantities (increase productivity) by resource saving or could increase output of solid waste collection with current input.

The ability of a company to obtain maximum output (100% technical efficiency) from given inputs depends on its operations management (capabilities) and economics of scale. There is evidence that variations in technical, pure technical and scale efficiencies in solid waste collection exist within and across the cities. There were variations in efficiencies within and across the cities. The influence of capacity (scale of operation and capabilities) on productivity is explored in this study by analyzing two groups of companies – those below and above average performance – using a criterion of industry average (50% technical efficiency) to split the companies. The differences in service operations and scale between the two groups of companies (below average performers and above average performers) will be explored. These differences will bring to the forefront the issue of capacity of the companies.

Results of Productivity using Ratio Analysis

The ratio analysis of productivity was conducted to further explain the inefficiencies obtained from data envelopment analysis. Vehicle productivity is defined as the ratio of output to input (actual output per vehicle per day). Figure 6.3 shows the vehicle productivity of all the 33 companies. The technical efficiency results from the data envelopment analysis and the productivity values from the ratio analysis are highly correlated (R^2=0.99) (Figure 6.2). The technical efficiency results and the productivity values showed similar pattern (Figures 6.1 and 6.3). The average productivity was 18 tonnes per day per vehicle and 14 out of 33 companies had values above this average. In Accra, 8 out of the 12 companies had productivity values above average. In Kumasi and Tema, 3 companies in each city had productivities above average. In Kumasi, companies C19, C20 and C23 performed above average. The performance of the private companies revealed that individual vehicle productivity varied across companies.

Whereas some companies in Accra, Kumasi and Tema performed above average, all the companies in Takoradi and Tamale performed below average. The vehicle productivity in Tamale was as low as 2 tonnes per day per vehicle compared to the performance of companies in the other cities. In Tamale the company with the highest vehicle productivity (C25) had vehicle productivity of 18 tonnes per day per vehicle which was the average productivity. Apart from C25 which had four collection vehicles, the rest of the companies had one collection vehicle each. With regard to the type of collection vehicles used, 4 of the low performing companies used conventional vehicles (farm tractors) not meant for waste collections while one of them used a compactor truck.

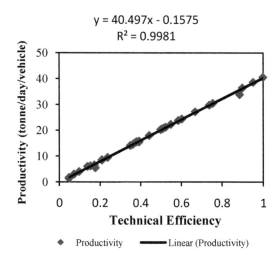

y = 40.497x - 0.1575
R² = 0.9981

Figure 6.2 Relationship between productivity and technical efficiency

The results show that there were variations in vehicle productivity within the same city and across cities. This study explores why all companies in Sekondi-Takoradi and Tamale performed below average, whiles some companies in Accra, Kumasi and Tema performed above average. Those companies below average may have common characteristics which will be explored in this study. The possible factors for low productivity values and underperformance to be explored are the company size, condition of vehicles, economics of scale, size of collection zone and waste generation, and capabilities for operations management. The size of the company (number of collection vehicles) to some extent, the type of vehicles and their conditions and the utilisation of vehicles (quantity of waste collected) as well as waste quantities generated could be the factors that accounted for the low performance. These factors will be further explored in the next chapter. Productivity depends on how the companies organise their vehicle operations to utilise the vehicles to make more trips in a day and take full vehicle load. The maximum productivity is achieved through vehicle utilisation by ensuring that each vehicle load is full and maximum number of trips for each day is achieved.

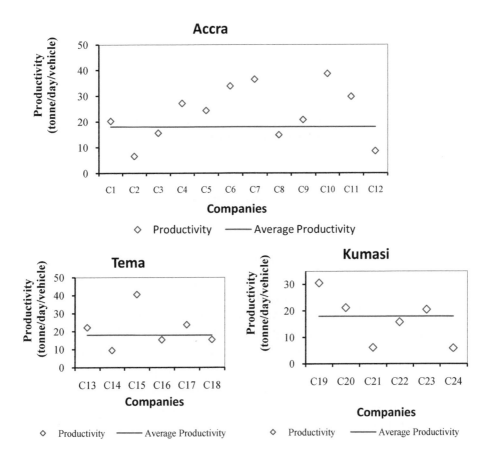

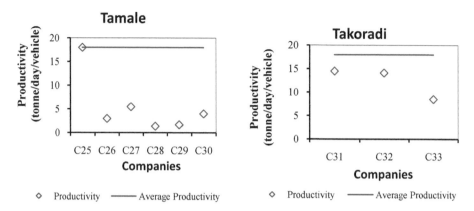

Figure 6.3 Productivity of companies in the cities, showing companies above and below average productivity

6.4   Factors Influencing Productivity and Efficiency of Companies

The companies have different vehicle types and varying number of vehicles. The utilization of vehicles of the same type differs from company to company. A vehicle which is highly utilised means that it makes the maximum daily number of solid waste collection trips and takes full load (full vehicle capacity) for each trip to the disposal site. Vehicle utilisation is a measure of how the various companies utilise their waste collection vehicles. High utilisation of vehicles leads to high productivity. Vehicle utilisation is an efficiency measure and is defined as the ratio of the output (actual waste collected by each vehicle) to the maximum capacity of the input (vehicle). In other words, vehicle utilisation measures the actual output of vehicles as against the maximum vehicle capacity (tonne per day). Table 6.2 provides the basis for the analysis of utilisation of the various types of vehicles used for solid waste collection by the companies. The data were obtained from analysis of monitoring of solid waste collection operations. The assumption for the computations of utilisation is that the maximum capacity (tonnes per day) is achieved when the vehicle load for each trip is full and that the maximum number of trips per day is made. This assumption provides the basis for the analysis of the utilisation.

Table 6.2 Daily maximum capacities of solid waste collection vehicles

City	Trips per day	Full load per trip (tonne)[3]	Maximum capacity per day (tonne)
Compactor truck[1]	2	10	20
Side loader[1]	2	5	10
Skip truck[2]	7	4	28
Roll-on truck[2]	7	9	63
Tipper truck[2]	7	9	63
Tipper truck (small)[2]	7	4	28
Tractor-trailer[1]	2	3	6

Source: Analysis of data collected by Author,
[1]*Vehicles are used for house-to-house collection service.*
[2]*Vehicles are used for communal collection service*
[3]*Full load per trip was obtained from previous study and was assumed to be the output of fully utilised solid waste collection vehicle.*

The number of trips and waste collected by each vehicle were used to determine the actual output for all the companies. The maximum capacity was computed using the optimum maximum number of trips each type of vehicle could make based on historical records and the nominal quantities of waste that could be disposed of by the various types of collection vehicles. For example, a compactor truck with nominal weight of 10 tonnes per trip could make two round trips a day and therefore the maximum capacity would be 20 tonnes per day. Vehicle capacity utilization gives indication of the extent to which the companies make use of their waste collection vehicles. Figure 6.4 shows the vehicle utilisation for the 33 companies in the five cities. The results show that the average utilisation value of all cities was 59% and 14 companies had values above this average.

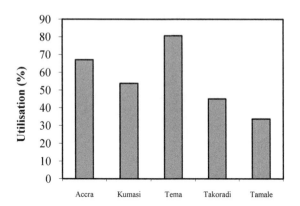

Figure 6.4 Average vehicle utilization of companies in the cities

In Accra, the average vehicle utilisation value of the 12 companies was 67% and 6 companies performed above all cities average of 59%. In Kumasi, the average vehicle utilisation of the 6 companies was 54% and only two companies (C19 and C20) utilized their vehicles and performed above the all cities average. Four companies in Kumasi had values below average. In Tema, the average vehicle utilisation of the companies was highest compared to averages of the other cities (Figure 6.4). The average vehicle utilisation value for the 6 companies in Tema was 81% and 5 companies were above the all cities average of 59%. In Takoradi, the average vehicles utilisation was 45% and all the companies were below all cities average. In Tamale, the results showed that the average vehicles utilisation of the 6 companies in the city was 34% and all the companies performed below all cities average.

Vehicles utilisation in Tema and Accra was higher than the other cities. The results indicated that vehicle utilisation varied across and within cities. One of the reasons for the high vehicle utilisation in Tema may be attributed to the planned road network in the city, with less traffic congestion compared to Accra and Kumasi. Another reason for performing above average may be good conditions of vehicles and the ability of the companies to manage their vehicles to achieve full vehicle load and maximum trips. The number of daily trips of vehicles in good conditions is high and this results in high vehicle utilisation.

Companies in Takoradi and Tamale performed below average. The reasons for the low average values of vehicles utilisation in Tamale and Sekondi-Takoradi and below average performers in the other cities may be due to sub-optimal scale of operation and management capability problems. On the whole, the vehicle utilisation of the companies in Tamale was low compared to companies in the other four cities. One of the companies had 4 collection vehicles on the road while the other 5 companies had one each. Four of the companies in Tamale had just one tractor each for solid waste collection, which could easily breakdown, and thus affect the vehicle utilisation. The scale of operation and management capabilities may explain the low utilisation.

Vehicle utilization positively correlates (R^2=71%) with productivity and DEA technical efficiency (Figures 6.5). As productivity and technical efficiency of the companies increase, the utilization also increases. There are two operational factors which determine the utilization of company vehicles. The first is the availability of each company vehicle to make the maximum daily number of trips. A trip is a single round of waste collected by a vehicle from the collection point to the disposal point. Vehicle breakdowns in the course of the day and collection planning usually affect the daily number of trips. The second factor is the vehicle load that is taken to the final disposal site. Utilising full vehicle capacity by taking full vehicle load of waste for each trip to the disposal site increases the productivity and hence the utilisation of the vehicles. There are three reasons for low productivity as well as low utilization in the cities. These are:

o Vehicle availability for 8 hours working day may not be achieved due to vehicles breakdowns in the course of the day or for some period of time.

o Low daily workload of vehicles due to few populations served or small service area and/or less economics of scale, arising from market concentration where greater proportion of the waste is collected by few large companies.

o Few daily trips and low workload due to inefficient operations management.

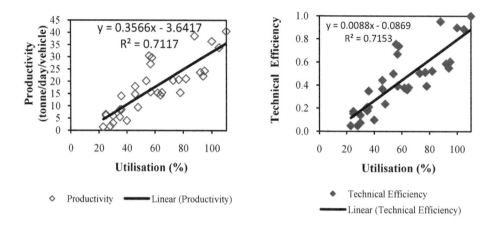

Figure 6.5 Variations in productivity and technical efficiency with utilization

Table 6.3 shows the operational characteristics of communal and house-to-house collection vehicles for companies with productivity above and below average. The vehicles used for communal collection were skip and roll-on trucks whereas compactor trucks were used for house-to-house collection. It can be noted that compactor vehicles used for house-to-house collection made two daily trips for companies above average in Kumasi and Accra but made four daily trips for companies above average in Tema. Apart from Kumasi, the companies below average had lower number of trips per day in Accra and Tema. For communal collection, the average number of trips per day was higher for companies whose productivity was above average than those companies whose productivity was below average in both cities.

Table 6.3 Operations output of companies above and below average performance

Operational Factors	Communal Collection vehicles		House-to-house Collection vehicles	
	Below average performers	Above average performers	Below average performers	Above average performers
Accra				
Average vehicle trips per day	2.6	5.9	1.3	2.2
Average vehicle load (tonne) per trip	6.5	7.2	7.0	9.9
Average vehicle load (tonne) per day	16.9	35.9	10.8	22.5
Kumasi				
Average vehicle trips per day	3.8	5.8	2.0	2.0
Average vehicle load (tonne) per trip	4.0	4.2	7.0	7.2
Average vehicle load (tonne) per day	14.0	25.0	10.0	12.5
Tema				
Average vehicle trips per day	2.3	6.8	1.7	4.1
Vehicle load (tonne) per trip	4.0	6.7	9.5	7.8
Average vehicle load (tonne) per day	9.2	33.3	15.8	28.3
Takoradi				
Average vehicle trips per day	2.9		1.4	
Average vehicle load (tonne) per trip	4.1		8.8	
Average vehicle load (tonne) per day	12.0		12.6	
Tamale				
Average vehicle trips per day	5		1.2	
Average vehicle load (tonne) per trip	3.6		2.9	
Average vehicle load (tonne) per day	18.1		3.1	

Source: Results of analysis of field data collected

There were variations in average quantity of waste collected (tonnes per trip) for both the compactor trucks and the skip trucks for companies whose productivity was above and below average. The quantity of waste collected (tonnes per trip) for the compactor trucks was high for above average performers than below average performers. Companies with vehicle productivity above average had higher average daily quantity of waste collected (in tonnes per vehicle) for both collection methods than those companies with productivity below average. However, the quantity of waste collected per trip for compactor trucks was lower than the maximum output of 10 tonnes per trip.

The actual output of vehicles of the companies below average was far below the maximum capacity, and hence the utilisation of vehicles by such companies was low. The factors which explain the difference in vehicles' utilisation and productivity by the individual companies were the number of daily trips and vehicle load (tonnes of waste collected per trip). Vehicle productivity of a given company is directly influenced by the number of trips that each collection vehicle makes per day and the quantity of waste collected in tonnes per trip as well as the daily quantity of waste collected (tonnes) by the individual vehicles. The number of daily trips is affected by the availability of vehicles on road since a vehicle may not be able to make the maximum trips for the day due to breakdown in the course of the day's operation. Another factor which influences utilisation is the management capability for collection planning and supervision. The low utilisation may be attributed to how route planning and vehicles operations are organised. If a waste collection company is able to maximise these factors, the utilisation and productivity of that company will increase. The combined effects of the number of trips that each collection vehicle makes per day, the quantity of waste collected in tonnes per trip and the daily quantity of waste collected (tonnes) by individual vehicles determine the level of a company's overall vehicles productivity.

There were differences in the daily number of trips and the quantities (tonnes) of waste collected for each trip as well as daily quantity of waste collected (tonnes) by vehicles. The low productivity values recorded by the companies in these cities could be due scale problems and management capability. An inefficient management of vehicle operations could reduce productivity of the vehicles. The companies whose productivity values were below average could not maximise the number of trips per day and tonnes of waste collected per trip. Increasing productivity depends on how the companies organise their operations to make optimum use of the vehicles in terms of the number of trips made in a day, tonnes of waste collected per trip and the quantity of waste collected in tonnes per day per vehicle.

The companies in Tamale and Takoradi were underperforming (low values of utilisation) and this may be attributed to vehicle conditions, vehicle breakdowns and weak management capability. The conditions of vehicles and internal management capabilities (vehicles supervision operations, route planning and vehicle maintenance) may account for the high utilisation values in Accra, Tema and Kumasi. Companies with high average values of vehicle productivity maximised the number of trips and quantity of waste collected by the various waste collection vehicles used. Achieving maximum output of waste collection vehicles depends primarily on how the collection companies organise their operation management. There were differences in the daily number of trips and the quantities (tonnes) of waste collected for each trip as well as daily quantity of waste collected (tonnes) by both the compactor trucks and the skip trucks for companies whose productivities were above and below average. The low productivity values recorded by the companies in these cities were attributed to scale problems and inadequate management capability.

The reasons for the high vehicle utilisation in Tema may be as a result of the planned road network in the city, with less traffic congestion compared to Accra and Kumasi. This allows the companies the ability to manage their vehicles to achieve full vehicle load and maximum trips. The vehicle utilisation of the companies in Tamale and Takoradi was lower compared to companies in the other cities. The low vehicles' productivity in Tamale and Takoradi and the companies below average performance in the other cities is due to sub-optimal scale of operation and lower vehicle utilization.

6.5   Service Quality of Companies

The quality of waste collection service was assessed in two ways. First, the residents were asked to rate companies (on a five-point scale from very poor to very good) in terms of three service quality attributes (reliability of collection, waste overflow and sanitary conditions at bin/container locations, and response to customer complaints). Second, the residents were asked to indicate service satisfaction by responding either satisfied or not satisfied with the quality of service. In customer satisfaction survey, service satisfaction is the customers' perceptions of service received (perceived quality) compared to their expectations of the desired service quality. When expectations are exceeded, service is perceived to be of exceptional quality. When expectations are not met, however, service quality is deemed unacceptable or poor. The service quality is satisfactory or good when perceived service quality meet expectations. The following section presents service quality results obtained from residents survey conducted in the five cities and discusses the differences in service quality across companies and cities.

Quality of Service of Companies in Accra

Residents were asked to rate service quality of private companies in their communities. Residential areas being served by all the companies were selected for the survey. Table 6.4 presents the results of residents' rating of two service quality attributes (reliability of collection, and waste overflow and sanitary conditions at bin/container locations). The survey shows that there was significant difference (Chi-square value = 187.7; df = 22; p<0.01) in the quality of service rendered by the companies in Accra.

The service reliability and sanitary conditions of house-to-house collection service in the high and medium income areas (served by C1, C2, C6, C8, and C12) were generally rated as fair and good by residents. Residents in Airport Residential Area (C1) and Cantonments (C6) which are high and medium income areas respectively rated the service quality of the private companies in their area as either fair or good. Sixty-seven percent (67%) of residents in Airport Area Residential indicated that quality of service was good whilst 76% indicated fair for Cantonments. The level of service quality could be attributed to the perceived quality of service by the residents, since these residents pay user charges that recover full cost and therefore expect the service to be reliable and devoid of waste overflow from bins located in front of their houses.

Table 6.4 Customers' rating of service quality of private companies in Accra

Company	N	Sanitary conditions and spill-over of waste (%)	Reliability of waste collection (%)
C1	33	68	67
C2	36	72	69
C3	18	53	49
C4	31	73	59
C5	15	60	56
C6	33	72	64
C7	19	57	56
C8	32	73	64
C9	27	62	63
C10	35	42	37
C12	20	60	59

Notes: Chi- Square value = 187.7, Critical Chi- Square value = 33.9, Sig. (2-sided) = 0.000; Level of Sig. < 0.01; 12 cells (36%) have expected count of less than 5. For contingency table more than two by two, expected frequency of 1 in 5 cells is allowed to be less than 5 (Varkevisser et al., 2003:143).

The service reliability and sanitary conditions of communal collection in the low income areas (served by C3, C5, and C10) were generally rated as poor or fair by residents. Interestingly, about 56% (C3) and 94% (C10) of residents in low income areas rated the quality of service as poor. This is because the waste spillover from communal containers unto the ground is common at most sites. The households confirm that collection is irregular and the containers' sites are not cleaned by the companies. Companies C2 and C8 were below average performance in terms of productivity and technical efficiency, but had service quality scores comparable with those of above average performance. This suggests that there is no correlation between service quality and technical efficiency. The aggregate service quality scores of below average performers (C2, C3, C8, C12) for reliability and waste overflow ranged from 49% to 73%, which are comparable with those of above average performance which ranged from 37% to 73%.

Service quality in Kumasi

Table 6.5 presents the results of residents' rating of service quality in terms of reliability of collection, and waste overflow and sanitary conditions at bin/container locations. The survey shows that there was significant difference (Chi-square value = 82.8; df = 10; p<0.01) in quality of service rendered by the companies. The service reliability and sanitary conditions of both house-to-house and communal collection services rendered by companies were generally rated as poor, fair and good by residents.

Table 6.5 Customers' rating of service quality of private companies in Kumasi

Company	N	Sanitary conditions and spill-over of waste (%)	Reliability of waste collection (%)
C19	67	45	51
C20	68	63	58
C21	58	58	63
C22	19	61	58
C23	45	35	48
C24	30	52	31

Notes: Chi- Square value = 82.8, Critical Chi- Square value = 18.3, Sig. (2-sided) = 0.000; Level of Sig. < 0.01; 7 cells (39%) have expected count of less than 5.

A single aggregate score was calculated for each company in order to compare the companies. A service quality aggregate score of 80% indicates that the company service quality is good while 40% or less indicates that the service quality of the company is poor. The service reliability aggregate scores of the companies ranged from 31% to 63%, indicating that most of the companies were rated as poor or fair.

The aggregate scores of the attribute of sanitary conditions and spill-over of waste at bin/container locations ranged from 35% to 63%. Reliability of collection and waste overflow are related to some extent. Waste spill over from bins and containers occurs when bins and containers get full and are not emptied or picked-up by the companies due to irregular collection.

Waste spill over from bins occurs as a result of unplanned waste collection or changes in waste pick-up scheduling and irregular pick-ups. Spill over of waste from bins in front of houses (house-to-house collection) and at container locations (communal collection) contributes to the poor aesthetic conditions and nuisance in communities even though companies may be operating in those communities.

Service quality in Tema

Table 6.6 shows the results of residents' rating of service quality in terms of reliability of collection, and waste overflow and sanitary conditions at bin/container locations in Tema. The survey shows that there was significant difference (Chi-square value of 39.5; df =8; p<0.05) in quality of service rendered by the companies in Tema.

Table 6.6 Customers' rating of service quality of private companies in Tema

Company	N	Sanitary conditions and spill-over of waste (%)	Reliability of waste collection (%)
C13	42	48	46
C14	41	59	45
C15	10	70	70
C16	31	45	39
C17	10	60	62

Most of the residents in the high income areas where house-to-house service is rendered (by C13, C14, C16) generally rated the service quality as poor or fair. In the high income communities served by C13 about 62% of residents rated the service quality as poor and 29% of the residents said it was fair. In communities served by C14 about 72% of residents said that the service quality was poor while 22% said it was fair. In communities served by C16, 84% said the service quality was very poor and 16% said it was fair. The service quality was perceived as poor or fair by most of the high income residents in the three high income zones. All these zones received house-to-house collection service and residents pay user charges that recover full cost of service. The variation in the perceived service quality (poor and fair) by the residents may be attributed to the difference in their desired expectations. A house-to-house collection service with comparatively reliable service without interruption in the collection frequency and pick-up days is usually rated as good. But this may not be the situation in Tema given the levels of responses as poor or fair in the three zones.

Companies C14 and C16 are below average performance and also had low aggregate scores for reliability and waste spillover (Table 6.7). However, C13 had productivity and technical efficiency above average but low aggregate scores for reliability and waste spillover. This indicates that high productivity and technical efficiency do not necessary lead to high service quality as theory suggest. High performance of a company is achieved when both productivity and service quality are high. The residents' perceptions in low income areas where communal collection is rendered by C15 and C17 indicate generally high level of quality of service. Most of the residents (75%) served by C15 rated the service quality as good. For C17, about 70 of the respondents said reliability was fair while 20% said it was good. Both C15 and C17 are above average performance in terms of productivity and technical efficiency.

Quality of Service in Takoradi

Table 6.7 shows the results of residents' rating of reliability of collection, and waste overflow from bin/container and sanitary conditions. The survey shows that there was significant difference (Chi- square value of 42.0; df = 4; p<0.05) in quality of service rendered by the companies. The service quality of Company C31 was higher than that of C32 and C33. All the three companies in Takoradi are below average performers in terms of productivity and technical efficiency. However, the aggregate service quality scores for reliability and waste overflow ranged from 50% to 71%, which are comparable with those of above average performance in the other cities.

Table 6.7 Customers' rating of service quality of private companies in Takoradi

Company	N	Sanitary conditions and spill-over of waste (%)	Reliability of waste collection (%)
C31	68	65	71
C32	90	53	54
C33	22	50	50

Service Quality in Tamale

Table 6.8 shows the results of residents' rating of reliability of collection, and waste overflow from bin/container and sanitary conditions. The survey shows that there was significant difference (Chi- square value of 19.3; df=10; p<0.05) in quality of service rendered by the companies. All the companies in Tamale are below average performance in terms of productivity and technical efficiency. However, the aggregate service quality scores for reliability and waste overflow ranged from 64% to 90%, which are comparable with those of above average performers in the other cities.

Table 6.8 Customers' rating of service quality of private companies in Tamale

Company	N	Sanitary conditions and spill-over of waste (%)	Reliability of waste collection (%)
C25	16	59	64
C26	13	74	75
C27	9	73	62
C28	10	96	96
C29	12	90	90
C30	10	67	87

Comparison of Performance of Companies above and below Average Performance

Figure 6.6 shows the relationship between service quality and technical efficiency of companies across the cities for companies below and those above average performance. There is no significant difference (t= 1.6; df=30; p>0.05) in the service quality of below and above average performers. For instance the service quality of four of the below average performers in Tamale was higher than the above average performers in the other cities. These companies in Tamale had high service quality scores because they served few house-to-house service customers and do not render communal service. These companies served between 140 to 520 customers compared to between 3000 and 8000 customers being served by the other companies in the other cities. The service demands (few customers) served by the companies were below the capacity of the companies, which means that their productivity could be increased by increasing the number of customers.

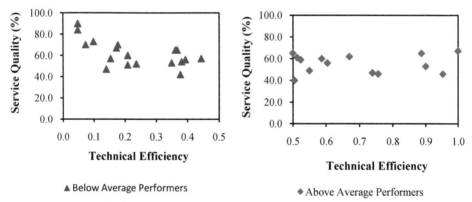

Figure 6.6 Relationship between service quality and technical efficiency of above and below average performers across the cities

The differences in performance between the two groups of companies (those below and those above the average) were also analysed using independent samples test. Table 6.9 shows a summary of independent samples test comparing the two groups of performers. The t-statistics (equality of means) provide a strong evidence of significant differences in productivity, technical efficiency and vehicle utilisation between the two groups, but there was no significant difference in service quality between the groups. The service quality of the companies was mixed, and also there was no significant correlation between productivity and service quality.

Table 6.9 Comparison of performance of the two groups of performers

Variables	t	df	Sig.	Mean Difference	Std. Error Difference
Service Quality	-0.207	21.99	0.838	-0.72	3.48
Productivity	6.240	22	0.000	18.04	2.89
Utilisation	8.237	22	0.000	52.88	6.42
Scale Efficiency	4.176	17.43	0.001	0.26	0.06
Technical efficiency	4.992	6.40	0.002	0.45	0.09

This analysis suggests that high productivity and technical efficiency compromise service quality if the capacities of the companies cannot cope with the increasing waste generation. There is no significant difference in the service quality of below and above average performers. For instance the service quality of four of the below average performers in Tamale was higher than the above average performers in the other cities. These companies in Tamale had high service quality scores because they served few house-to-house service customers and do not render communal service. The service demands (customers) served by the companies were below the capacities of the companies, which means that their productivity could be increased by increasing the number of customers. Some of the companies with technical efficiency and productivity above average had low aggregate scores for reliability and waste spill-over. Five small companies in Tamale had high service quality due to fewer population served, but their scales of operation were inefficient, resulting in low productivity.

In theory, a well performing company achieves high productivity and provides better service quality. But the study suggests that the inefficient use of resources account for the low productivity and poor service quality of the companies. Adequate private sector capacity for service provision ensures that resources are used more efficiently and also stimulates innovations and competitive advantage to achieve high productivity and provide better service quality. This analysis suggests that high productivity and technical efficiency do not necessarily lead to higher service quality. High performance of a company is achieved when both productivity and service quality are high.

6.6   Factors Influencing Service Quality of Companies

Table 6.10 shows the results of service quality across the cities. Chi square analysis shows that there was significant difference (Chi-square value = 340; df = 8; p<0.05) in quality of service across the cities. Residents' perceived service quality across the five cities shows that service quality for Tamale (74.4%) was the highest, followed by Accra (61%). The service quality ratings for Kumasi, Tema, and Takoradi were 51%, 50% and 56% respectively.

Table 6.10 Results of service quality scores across the cities

City	Sample size (N)	Reliability (%)	Sanitary conditions and waste overflow (%)	Quick response to complaints (%)	Overall service quality (%)
Kumasi	279	52.2	50.61	47.5	50.1
Accra	459	60.2	64.32	58.2	60.9
Tema	181	46.6	53.49	50.6	50.2
Takoradi	255	58.2	56.63	54.4	56.4
Tamale	104	80.2	78.27	64.5	74.4

Solid waste collection is a visible service and residents are better able to rate the perceived service quality. The level of service (house-to-house and communal), subsidies and payment of user charges could influence the perceived service quality observed across the cities.

Table 6.11 shows the results of service quality for both collection methods. Residents' perceived service quality for the two methods of collection across the five cities showed that service quality for house-to-house collection was slightly higher than communal collection. Some of the respondents served with house-to-house collection indicated that their waste containers did not overflow because the collection was frequent as scheduled. In communities where there was regular collection of waste generated, the residents' rating of service quality was high, whereas in communities where collection was irregular the rating was low. Service quality for house-to-house was high because residents pay for the collection service and as a result, for some of the companies, the collection was regular as scheduled and reliability of companies was high.

Table 6.11 Scores of reliability and waste overflow of house-to-house and communal collection services

| | Reliability of waste collection | | Sanitary conditions and spill-over of waste at bin/container site | |
City	House-to-house collection (%)	Communal collection (%)	House-to-house collection (%)	Communal collection (%)
Accra	63.8	49.3	68.4	52.1
Kumasi	58.8	45.8	64.8	38.6
Tema	45.8	50.0	53.8	50.0
Takoradi	67.8	48.5	63.8	47.6
Tamale	79.0	70.5	78.0	68.4

Table 6.12 presents the level of satisfaction of residents. Chi-square analysis shows that there was significant difference in the levels of satis-

faction across the five cities for communal collection and house-to-house service. In general, the user satisfaction for house-to-house collection was higher than that of communal collection. This result is consistent with finding from Awortwi (2003:209) that house-to-house service quality was higher than that of communal collection. House-to-house offers incentives to render better service since companies cannot afford to perform poorly because users will not honour their financial obligation. The desire of the users to get value for their money implies that the role of users in demanding accountability and responsibility from service providers come to the fore compared to the Assemblies' role in service quality monitoring. In Tema, the households' satisfaction for house-to-house service was higher than communal collection. Interestingly, only 24% of the residents indicated that they were satisfied with the quality of the house-to-house collection service. Similar finding was obtained by Awortwi (2003). This may probably be due to the fact that their desired expectation of service quality was not met.

Table 6.12 Rating of response to residents' complaints and overall satisfaction of house-to-house and communal collection services

City	Quick response to residents' complaints		Satisfaction (% respondents satisfied with service quality)	
	House-to-house collection (%)	Communal collection (%)	House-to-house collection (%)	Communal collection (%)
Accra	59.0	56.1	70.7	43.9
Kumasi	54.9	40.4	53.4	11.6
Tema	51.0	47.8	23.9	56.0
Takoradi	58.7	53.1	91.6	21.2
Tamale	67.2	63.3	81.7	63.2

Satisfaction is the difference between the perceived quality and the expectation of the customers. Households may be satisfied if their perceived quality is the same as their expectation, and may be dissatisfied if their perceived quality is less than expectation. The satisfaction of service by residents receiving communal collection service may be different from those who pay for house-to-house collection. This is because the service cost of the communal collection is covered by subsidy borne by the Metropolitan Authorities and central government. The satisfaction for communal collection in Kumasi and Takoradi was low. Communal collection does not offer incentives to render better service (Awortwi, 2003:210), because the people did not pay for the service and subsidy payment delays most of the time.

6.7 Conclusions

In this chapter, productivity, DEA technical efficiency and service quality of private sector companies were analysed. The differences in performance were explored by analyzing two groups of companies – those below and those above average performance – using a criterion of industry average (50% technical efficiency) to split the companies. It was identified that there were variations in productivity, technical efficiency and service quality of companies within and across the five cities. The analysis of these performance variables shows that 19 out of the 33 companies had technical efficiency values below 50% and productivity values below an average of 18 tonnes/day/vehicle. The companies that were less productive with efficiency rating of less than 50% were underutilizing the vehicles. The reasons for the inefficiencies are attributed to inefficient operations and suboptimal scale of operation. The results show that 14 out of 33 companies had values above average productivity of 18 tonnes per day per vehicle. Whereas some companies in Accra, Kumasi and Tema performed above average, all the companies in Takoradi and Tamale performed below average. As productivity of the companies increases, the utilization also increases and verse-versa. The results show that the average utilisation value of all cities was 59% and 14 companies had values above this average. Vehicles utilisation in Tema and Accra was higher than the other cities. The results indicated that vehicle utilisation varied across cities.The analysis also suggests that there was no correlation between service quality and productivity.

There following four conclusions on the findings of productivity and service quality of the companies can be drawn. First, it is concluded that low productivity of companies arises due to the low utilization of vehicles and sub-optimal scales of the companies. Companies with vehicle productivity above average had higher average daily quantity of waste collected (in tonnes per vehicle) for both collection methods than those companies with productivity below average. The actual output of vehicles of the companies below average was lower than that of the above average performers. The factors which explain the difference in vehicles' utilisation and productivity by the individual companies were the number of daily trips and vehicle load (quantity of waste collected per trip). The number of daily trips is affected by the availability of vehicles on road since a vehicle may not be able to make the maximum trips for the day due to breakdown in the course of the day's operation.

Second, it can be concluded that there was no significant correlation between technical productivity and service quality of companies. The t-statistics provide a strong evidence of significant differences in productivity, technical efficiency and vehicle utilisation between the two groups, but there was no significant difference in service quality between the groups. The service quality of the companies was mixed, and also there was no significant correlation between productivity and service quality. This was contrary to the hypothesis that high productivity of a company leads to better service quality and that high performance is achieved when both productivity and service quality are high. This study shows that high productivity and technical efficiency do not necessarily lead to higher service quality.

Third, another conclusion is that there were significant differences in service quality across cities and that the service quality of house-to-house collection was higher than that of communal collection. Solid waste collection is a visible service and residents are better able to rate the perceived service quality. The level of service (house-to-house and communal), subsidies and payment of user charges could influence the perceived service quality observed across the cities. Service quality for house-to-house was high because residents pay for the collection service and as a result, for some of the companies, the collection was regular as scheduled and reliability of companies was high. This result is consistent with finding from Awortwi (2003:209) that house-to-house service quality was higher than that of communal collection. House-to-house offers incentives to render better service since companies cannot afford to perform poorly because users will not honour their financial obligation. Communal collection does not offer incentives for better service because the people did not pay for the service and subsidy payment delays most of the time. The desire of the users to get value for their money implies that the role of users in demanding accountability and responsibility from service providers come to the fore compared to the Assemblies' role in service quality monitoring.

Finally, it can be concluded that the findings of this study is contrary to the hypothesis that high productivity of a company leads to better service quality, and high performance is achieved when both productivity and service quality are high. The difference in performance bring to the fore the issues of the capacity of the companies. The influence of capacity (scale of operation and capabilities) and regulation practices on performance of the private sector (productivity and service quality) will be explored in chapters 7 and 8 respectively.

7   Private Sector Capacity: Scale of Operations and Capability of Companies

7.1   Introduction

In chapter 6, productivity, DEA technical efficiency and service quality of private sector companies were presented. The analysis of these performance variables shows that 19 out of the 33 companies had technical efficiency values below 50% and productivity values below an average of 18 tonnes/day/vehicle. It was identified that there were variations in the productivity and technical efficiency of the companies across the cities. However, there was no significant correlation between technical efficiency and service quality. This suggests that high productivity and technical efficiency do not necessarily lead to high service quality. This chapter will examine the reasons for these relationships, but the analysis will focus more on technical efficiency and productivity which are objective variables compared to service quality which is subjective as it was based on costumer survey.

This chapter examines the relationships between performance and company capacity (scale of operation and capabilities). The research questions this chapter seeks to explore are: Does scale of operation of a company in solid waste collection influence performance? Do private sector companies have the capabilities and skills for operations management? The hypotheses to be tested in this chapter relate to these research questions. The hypotheses are that high scale of operation of a company leads to high productivity, and high capability of a company leads to high productivity. The variation in performance of companies that is explained by the capacity of the companies will be explored. The capacity of companies as used in this study means the scale of operation and the capability of companies to perform operations management tasks. The number of company vehicles and market share of waste collected by the companies were used as proxy for the scale of operation. The company capability is the ability of the company to perform operation management tasks involving solid waste collection planning, operations supervision and maintenance management.

Besides testing the hypotheses, the reasons for the differences in performance between two groups of companies – those below and those above average performances – will be explored in this chapter. Statistical analysis using independent sample tests and multiple regressions will be used for comparing the groups and testing the hypotheses respectively.

This chapter is organised into four sections. Section one gives a brief introduction of the chapter, while section two explains the influence of scale of operation on performance. Section three explains the influence of company capabilities on performance. The last section draws conclusion from the analyses on the relationships between performance of the companies and their capacities.

7.2   Influence of Scale of Operation on Performance

Scale Inefficiencies Depicted by DEA

Figure 7.1 shows the relationships between market share and DEA scale efficiency results for the 33 solid waste companies. Scale efficiency measures the inefficiency of a company which is attributed to scale of operation or disadvantageous conditions arising from the zone of operation. This advantageous conditions may be low demand for the service or few households served in the given solid waste collection zone of operation.

There was diminishing return to increasing scale of operations beyond the optimum scale. Company C15 has 7 vehicles and is the Most Productive Scale Size (MPSS), where there is constant return to scale. Twenty-five of the companies had increasing return to scale (IRS) while 7 had decreasing return to scale. A company can be technically efficient but the scale of operation may not be optimal, and as such the company may be performing with increasing returns to scale or decreasing returns to scale. For example, 26 companies which were too small in their scale of operation had increasing returns to scale (IRS). Seven companies which were large in their scale of operation had decreasing returns to scale (DRS). For both IRS and DRS, the efficiencies of the companies can be improved by changing their scale of operations. Scale efficiency measure indicates the amount by which productivity can be increased by moving to the point of technically optimal productive scale. The most productive scale size is the point on the production frontier where there is neither IRS nor DRS.

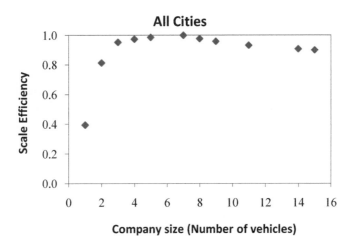

Figure 7.1 Relationship between DEA scale efficiency and company size for all 33 companies (Source: analysis of waste disposal data by author)

There was decreasing return to scale beyond 7 vehicles. This study suggests that the optimum scale of operation is 7 vehicles. For the purpose of economics of scale, companies need to operate at optimum number of 7 vehicles.   Companies with one vehicle in operation were scale inefficient. For example, 5 companies in Tamale, and 3 companies in Kumasi, Tema and Accra which were operating one vehicle each were scale inefficient. These companies had scale efficiency of 40% or less. In the case of Tamale, the companies were under utilising their scale due to few households served. The five companies were given one zone which is a size usually assigned to one company for economics of scale reason. The scale efficiency could improve if these companies are given a zone each instead of five companies operating in one zone. The inefficiencies of the companies is attributed to the fact that the scales of the companies were higher than waste to be collected from areas of operation or the companies were not able to optimise full load and maximum vehicle trips of waste collection arising from inefficient operations.

   The Figure 7.1 shows that 20% of the companies had scale efficiencies below 80%. These companies were scale inefficient. The scale efficiency values increased from 34% for inefficient companies to 100% for the most efficient company and decreased again. The average company and the least scale efficient company could have collected output measure with only 80% and 40% of their inputs respectively. These companies had excess input resources which were not being used, and therefore were operating under capacity. It also means that the scale inefficient companies had low productivity (despite the excess input) due to inefficient use of input resources.

Influence of Scale of Operation on Productivity

The relationships between performance variables and capacity using statistical analysis were explored to find out whether scale matters in solid waste collection. Table 7.1 presents descriptive statistics and Pearson correlation matrix for the performance variables (productivity and utilisation) and some of the variables that affect the performance variables (company size, market share of waste collected). The company size and market share positively correlated with productivity and technical efficiency. Increasing scale of operation leads to increasing productivity. The study reveals that the scale of operation had influence on productivity of solid waste collection. There was significantly positive correlation (0.84; $p < 0.01$) between the productivity and utilisation. The productivity of companies depends on the extent of vehicle utilisation.

Table 7.1 Descriptive Statistics and Pearson Correlations of capacity and performance

		mean	S.D	1	2	3	4	5
1.	Productivity	17.4	10.99					
2.	Technical efficiency	.43	.27	.999[b]				
3.	Scale efficiency	.79	.23	.630[b]	.622[b]			
4.	Capacity utilisation	59.09	26.01	.844[b]	.846[b]	.529[b]		
5.	Number of vehicles	4.54	4.13	.441[a]	.432[a]	.514[b]	.238	
6.	Market share	10.59	11.35	.425[a]	.418[a]	.505[b]	.242	.475[b]

[a] *Correlation is significant at the 0.05 level (p<0.05),* [b] *Correlation is significant at the 0.01 level (p<0.01)*

The size of a company (number of vehicles) is positively correlated with market share (correlation of 0.47; $p < 0.01$), but does not significantly correlate with utilisation. The capacity utilisation does not correlate with the number of vehicles (correlation of 0.23; $p > 0.05$) and market share (correlation of 0.24; $p > 0.05$). The factors that may explain the differences in company size and utilisation are the size of population served and vehicle conditions. Increasing the number of vehicles owned (size of company) may not significantly increase utilisation since apart from increasing scale of operation other factors also contribute to utilization and improved productivity.

Table 7.2 shows the result of the regression analysis of the capacity variables (independent variables) on the productivity (dependent variable). The dependent variable in this regression analysis is productivity. In order to determine to which extent company capacity explains the variation in productivity, capacity variables alone were incorporated in the regression model. The reason for testing such a model is that the test can indicate whether the capacity variables explain the variation of the dependent variable. The R square for the model was statistically significant (F= 4.6; p<0.05) and it explained 29% of the variation in productivity. Market share significantly explained partly the variation in productivity. This model suggests that high scale of operation leads to high productivity.

Table 7.2 Results of influence of capacity on productivity

Variables	Unstd. B	std. error	Beta(β)	t-value	Sig.
Constant	9.047	2.377		3.806	.001
Market share	.480	.159	.542	3.028	.006
R^2	0.294				
R^2 Adj.d	0.262				
F value	9.17				
Sig. F	0.006				
Durbin-W	0.865				
VIF	1.0				
N	25				

[a]Significant, P < 0.05; [b]Significant, P < 0.01

Figure 7.2 shows the relationship between vehicle productivity and the market share. The market share is the ratio of the quantity of waste collected by a company to the total waste collected in the city. The average productivity was 18 tonnes per day per vehicle and seven companies had values below this average. The productivity increases with increasing market share but declines as the market share becomes large (Figure 7.2). This study suggests a non-linear relationship between productivity and market share. The analysis of performance of private companies showed that the productivity of a particular vehicle type varied from company to company.

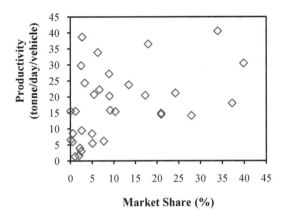

Figure 7.2 Scatter plot showing relationship between productivity and the market share

Figure 7.3 shows the relationship between vehicle productivity and the size of company using the number of vehicles as proxy. Thus, the number of vehicles on road was used as a measure of the size of company. The companies with productivity above the average of 18 tonnes per day per vehicle have varied number of vehicles (from 5 to 15 vehicles). Apart from few companies with one vehicle each which had high service quality values but low productivity, the companies with two or more vehicles did not show differences in service quality (Figure 7.3).

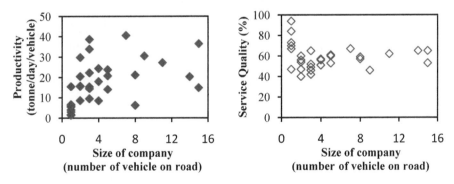

Figure 7.3 Scatter plot showing relationship between productivity and the number of vehicles on road

Figure 7.4 shows the relationship between vehicle utilisation and the market share and the number of vehicles. The results show that the average capacity utilisation value was 59% and 20 companies had values below this average. The capacity utilisation values increase with increasing market share, but some companies with market share exceeding 10% had low utilization values. The analysis suggests that companies with higher market share or more vehicles (beyond 7) have to manage the utilization of their vehicles in order to achieve higher productivity.

The utilisation values increase with increasing number of vehicles but some companies with vehicles exceeding 7 had low utilization values. The results suggest no correlation between the size of the company and the vehicle utilization, due to the low utilization of some companies with more vehicles. The decline in utilisation may be attributed to irrational route planning, discretional use of routes by drivers and how the companies organise their vehicles' operations.

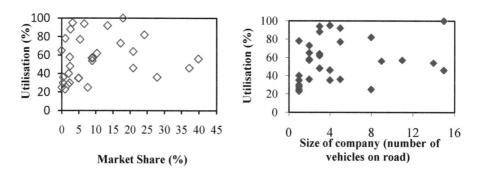

Figure 7.4 Scatter plot showing relationship between vehicle utilization and the company scale of input and output

The productivity of companies depends on how the companies organise their operations to utilise the vehicles in terms of the number of trips made in a day and the vehicle load. The maximum productivity is achieved by ensuring that each vehicle load is full and maximum number of trips per day is made. The size or scale of the company has relationship with productivity. It appears that there is certain threshold in terms of market share and company size above which productivity declines if prudent operations management is not done. However, certain small companies (below 10-15% market share) with less than 4 vehicles seem to have low productivity. To maximise productivity the companies have to plan their collection routes for the drivers; they do not have to allow the drivers to use their discretion as to which route to take.

7.3   Influence of Capability of Companies on Performance

The ability of a company to obtain maximum output (100% technical efficiency) from given input depends on its operations management (efficient operations) and scale of operation. The inefficiencies in scale of operation are measured by DEA scale efficiency whereas the inefficiencies in operations management are measured by DEA pure technical efficiency. Figure 7.5 shows results of the relationships between company size and DEA pure technical efficiency. There was diminishing return of pure technical efficiency to increasing scale of operations beyond the optimum scale. As the number of vehicles increases, the pure technical efficiency increases to a point and decreases again. A company with 7 vehicles has the optimum scale of operation. For a company with more vehicles beyond 7, the necessary capability is needed to manage better the fleet of vehicles in order to ensure high productivity.

Figure 7.5 shows there was diminishing return to increasing scale of operations in terms of market share beyond the optimum scale. The inefficiencies measured by pure technical efficiency are partly due to inefficient operations management within the companies. The inefficiencies due to inefficient operations cut across companies with low and high market shares of waste collected. The results showed that 7 out of 14 companies above average productivity performance had pure technical efficiencies of less than 70%. This means that there exist inefficiencies in the operations of these large companies which are above average performers and had market share ranging from 3.3% to 24%.

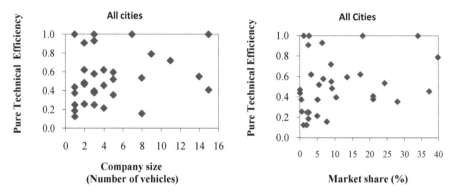

Figure 7.5 Relationship between DEA pure technical efficiency and company size for all 33 companies in the 5 cities (Source: analysis of waste disposal data by author)

It is evident that operations of certain companies below and above average were inefficient. The question is do these companies have the capabilities to perform operation management tasks such as solid waste collection planning, supervision and maintenance. The inefficiencies in operations management measured by DEA pure technical efficiency could be attributed to the differences in the capability of the companies alone but not disadvantageous conditions (size of service areas or scale) of the companies.

Table 7.3 presents the Pearson's correlation between the performance variables and the capability factors (independent variables). The management capability of private companies was assessed using their ability to perform operational tasks such as vehicle operations planning, operations supervision capacity and vehicle maintenance capacity. Supervisory capacity which measures whether technical personnel has the ability to supervise operation and whether top management is capable of managing the vehicle operations indicated positive and statistically significant correlation with vehicle productivity (Pearson's correlation of 0.398; $p<0.05$). This means that when supervisory capacity is rationalized (well planned and managed) vehicle productivity will increase. Depending on how vehicles operations are supervised, productivity may be increased or reduced.

Market share also shows positive and statistically significant correlation with vehicle productivity (Pearson's correlation of 0.423; $p<0.05$). This shows that when the market share is large, vehicles productivity may increase or otherwise. However, vehicle route planning, workers morale, vehicle condition and road condition indicated positive correlation but not statistically significant with vehicle productivity. Results of correlation between productivity and capability for vehicles route planning, operation scheduling, in-house maintenance were mixed and statistically insignificant.

Table 7.3 Correlation matrix of companies' capabilities and performance

Variables	1	2	3	4	5	6	7	8
1 Technical efficiency	1.00							
2 Productivity	0.99^b	1.00						
3. Years of experience	0.52^b	$.514^b$	1.00					
4. Route Planning	.149	.137	.093	1.00				
5 Operation Schedule	.009	-.007	.242	.258	1.00			
6 Supervision capacity	$.374^a$	$.369^a$	-.129	.240	-.266	1.00		
7 Workers morale	.099	.096	-.140	.077	-.234	$.323^c$	1.00	
8 In-house Maintenance	-.011	.003	.233	.167	$-.307^c$	.016	.204	1.00
9. Maintenance scheduled	-.164	-.164	-.070	.061	-.190	$.340^a$	.098	.203

^{a}*Significant, P < 0.05;* ^{b}*Significant, P < 0.01;* ^{c}*Significant, P < 0.1*

Table 7.4 shows the result of the regression analysis of the capability variables (independent variables) on the productivity (dependent variable). The dependent variable in this regression analysis is productivity. Two regression models were tested using all the capacity and capability variables (model 1) and then all variables that were not statistically significant were removed (model 2). In order to determine to which extent company capability explains the variation in productivity and technical efficiency, capacity and capability variables were incorporated in the regression model. The capacity variables were added to the regression model with the explicit purpose of explaining part of the variance. The reason for this analysis was to test whether the capability variables explain the variation of the performance variable.

Table 7.4 Results of influence of management capabilities on productivity

Variables	Model 1				Model 2			
	Unstd. B	Beta(β)	t-value	Sig.	Unstd. B	Beta(β)	t-value	Sig.
(Constant)	-26.98		-1.67	.116	-24.28		-2.168	.043
Years of experience	1.496	.699	5.719	.000[b]	1.393	.651	6.209	.000
Route Planning	-.032	-.046	-.356	.727				
Operation Schedule	.011	.016	.110	.914				
Supervision capacity	.327	.284	1.904	.078	.417	.362	3.133	.005
Workers morale	.123	.185	1.496	.157				
In-house Maintenance	-1.63	-.083	-.587	.566				
Maintenance schedule	-.125	-.184	-	.152	-.144	-.213	-1.923	.070
R^2	.834				0.801			
R^2 Adjusted	.727				0.759			
Sig. F change	000				0.142			
F value	7.8				19.08			
Sig. F (ANOVA)	0.00				0.00			
VIF	1.1-				1.1-1.9			
Durbin-Watson	2.51				2.24			
N	25				25			

[a]*Significant, $P < 0.05$;* [b]*Significant, $P < 0.01$*

The analysis shows that both models were statistically significant ($p < 0.01$) and that models 1 and 2 explained 83% and 80% (adjusted R square) of the variation in productivity respectively. The results of model 2 show that market share, years of experience of companies, supervision capacity, and to some extent, maintenance scheduling significantly explained partly the variations in productivity. This suggests that companies' experience, scale of operation, supervision capacity, and maintenance scheduling lead to high productivity. The results of the influence of route planning, operation scheduling, workers' morale and in-house maintenance on productivity of the companies were mixed.

Similar analysis using technical efficiency as dependent variable gave similar results, since productivity and technical efficiency are highly correlated. The analysis shows that both models were statistically significant ($p<0.01$) and that models 1 and 2 explained 83% and 80% (adjusted R square) of the variation in productivity respectively. This analysis shows that scale of operation and capability for operation management such as supervision capacity and maintenance scheduling lead to high productivity. It can be concluded that high scale of operation of a company leads to high productivity, and high capability of a company leads to high productivity. It can be said that scale of operation of a company matters in solid waste collection. However, certain private sector companies do not have adequate capability and skills necessary for some operations management tasks. The differences in capability of the companies were explored by analysing the tasks performed by the key operational management staff.

Vehicle Operations Planning Capability

Vehicle operations capability involves route planning and operation scheduling. Route planning is the detailed assignment of collection vehicles and labour to collection routes such that collection efficiency is maximized. Routing often includes the actual route that a vehicle should take to minimize either the distance that it covers or the time taken to complete the collection assignment. Table 7.5 indicates the responses of companies on the vehicle operations capability solid waste collection. It was asked whether the drivers were daily assessed to ensure that routes planned for them were strictly followed and that the assessment was done based on their daily output in terms of vehicles utilisation.

Out of twenty-five (25) companies, twenty-three (23) companies said that vehicles route plans were prepared for drivers to follow and that vehicles followed specific routes for solid waste collection and disposal. Vehicles route planning is necessary because it avoids waste in terms of fuel consumption. It also maximizes the use of time that contributes to an increase in the daily output of the vehicles. On expertise for vehicles route planning, ten (10) companies relied on external experts who are hired on part-time to plan their vehicles operations whereas 15 companies had in-house expertise to plan vehicles operations. It meant that those 15 companies had the required internal capacity to manage their vehicles operations, but their competences may be different.

Table 7.5 Responses of companies' managers on vehicle operations capability

Indicators	Disagreed (N)	Agreed (N)
Route Planning		
Vehicles route plans are prepared for drivers	2	23
Drivers follow specific planned routes	2	23
In-house expertise is available for analysis and planning of vehicle operations	15	10
Operation Scheduling		
Waste pick-up schedules are prepared and strictly followed	2	23
Drivers do not use their own discretion about routing	12	13

Source: Field Survey by Author

It was noted that 23 companies out of the 25 companies said that waste pick-up schedules were prepared and strictly followed while 2 companies did not strictly follow waste pick-up schedules. Waste pick-up schedule means certain days and times of the day are planned for pick-up of containers (communal collection) or emptying of bins (house-to-house service). Waste pick-up schedules are necessary to ensure that waste generated by the communities is picked-up as scheduled to avoid inconvenience such as spillover and odour nuisance at the collection points. It also determines the performance of the companies. Waste pick-up schedules if strictly followed contributes to productivity (waste collected in tonnes per day per vehicle) and quality of service of a company.

Another aspect of the operation scheduling was to find out whether drivers use their own discretion about vehicle scheduling and routing for waste collection. The survey showed that 13 companies said drivers did not use their own discretion about scheduling and routing. On the other hand, 12 companies said drivers did use their own discretion about scheduling and routing for waste collection. This means that some drivers do use their own discretion about scheduling of pick-ups of waste and which route to take. The drivers do not follow the roads planned for them but rather they use their own discretion about scheduling and routing. The decision of the drivers in this regard is sometimes due to odour nuisance when the waste is not covered especially when there is high traffic congestion along the planned road. To maximize daily output of waste collection depending on the prevailing conditions, drivers normally opt to use their own discretion about scheduling and routing to avoid traffic congestion.

One of the aspects of management of the solid waste collection is route planning and preparation of planned arterial roads for waste collection drivers to follow so that they can maximize the daily output of waste collection. It therefore means that if drivers do not use the planned roads their daily output may not be achieved and in addition fuel may be wasted thereby affecting the performance of the companies. One aspect of increasing vehicle productivity is to select and plan routes so that the quantity of waste collected daily is maximized. Often, task of choosing vehicle routes is assigned to the supervisor or left to the discretion of vehicle drivers.

Vehicle operation planning is necessary for solid waste collection since controlling and managing ahead of time ensures that full loads are optimized and partial load trips minimized. The use of route plans, vehicle log books for recording times and mileage and waste pick-up schedules impact positively on the performance of solid waste companies. From the observations made during the field study it was noted that some companies employed the use of route plans, log books and waste pick-up (tour) schedules whereas other companies did not. Strict adherence to these practices, in no doubt, aid the operations of the solid waste companies and enhances productivity. Subsequent analysis of log book data by supervisors ensures that improvements are made in terms of reliability and how work is carried out.

To further understand the companies' capability for operations planning, detailed analysis of actual practices in 13 selected companies were carried out by comparing above and below average performers. The results of capability of companies above and below average for operation plan-

ning were mixed (Table 7.6). This explains why the regression results
show that operations planning did not influence productivity.

Table 7.6 Vehicle operations planning capability of above and below average performers

Variables	Above Average Performers (N=6)		Below Average Performers (N=7)	
	Performed Task	Did not Performed Task	Performed Task	Did not Performed Task
Vehicle route plans are prepared	4	2	4	3
Drivers use vehicle route plans	4	2	4	3
Drivers use pick-up schedules	5	1	7	0
Waste pick-up schedules followed	3	3	6	1
Drivers do not use their own discretion on routes to take	4	2	4	3
Drivers follow specific routes for waste collection	4	2	4	3
Traffic studies are conducted	3	3	4	3

Companies which were more capable of planning operations performed a number of tasks. In such companies vehicle scheduling and route planning are done for the drivers. Companies with less capability did not have well defined routes for waste collection. The companies do not have any route plan prepared for their drivers so the drivers do use their own discretion as to the route they take. There are no plans to reduce delays as the companies do not have route map of the areas and alternate roads in case there is congestion on the main route. Besides, the waste pick up schedules are used but not strictly followed. The companies did not conduct any studies on traffic to avoid traffic delay. The study suggests mixed findings on the conduct of traffic studies to avoid delays, setting of output targets by management for vehicles, and minimization of the time spent by breakdown vehicles off the road.

Operations Supervision Capability

Supervision involves watching over an activity being carried out by some-body and ensuring that it is performed correctly. In solid waste collection, vehicle operations supervision may start with the allocation of a certain number of vehicles to a supervisor. The supervisor's work may then span from checking of records in vehicle log books, monitoring and follow-up of waste collection as well as keeping records of the daily number of trips and tonnage of waste. For some organizations, management may set output targets for each vehicle and hence the supervisor's job may include ensuring that these targets are achieved.

The results show that all the companies said technical personnel had the ability to supervise vehicles operations (Table 7.7). Although the re-sponsibilities of supervisors in the various companies were more or less similar, there were some cases in which duties differed slightly. Technical personnel are those personnel who have the technical know-how to super-vise vehicle operations. There were differences in the qualifications and salaries across the companies for all who hold supervisory position. The qualifications of supervisors ranged from secondary to tertiary. The pro-ductivity and utilisation of the waste collection vehicles depend on the ex-tent to which operations are supervised.

Table 7.7 Operations supervision capacity

Indicators	Disagree	Agree
Supervisory Capacity		
Technical personnel has ability to supervise operations, monitor and follow-up	0	25
Management is capable of managing the vehicle operations and analyse log books	0	25
Workers' Morale		
High morale of the employees due to fair labour practices	4	21
Collection crew put on protective clothing	1	24

Source: Field Survey by Author

Alongside the supervision of vehicles operations by the technical personnel is the role played by the top management of the companies. Top management in this context means directors and waste managers of the company who see to the day-to-day operations of the solid waste collection business. All the companies said top management was capable of managing the vehicle operations. There were differences in the qualifications and salaries across the companies for all who hold manager and supervisor positions. The managers' qualifications ranged from certificate, diploma to BSc and MSc. The degree of management effort therefore, clearly indicates the differences in performance of the various companies. The study was to find out whether the top management was capable of managing the vehicle operations. Management plays a role in the growth of every organization. In this regard, the quality of service, the productivity and utilisation of vehicles of the companies involved in solid waste collection are influenced by management to some extent.

Another aspect that determines the performance of the companies is the morale of the workers. In this domain, the study sought to find out if the morale of the employees was high due to fair labour practices and as to whether drivers and collection crew put on protective clothing during waste collection. Workers morale and motivation is an aspect of organisation that influences productivity. The responses indicated that 24 companies use protective clothing. The drivers and the crew members wore protective clothing during waste collection and that the dress code was meant to identify the drivers and crew members for each waste collection company. As part of making people aware of the value of solid waste management and the efforts of the service providers, workers in the solid waste system need clean and brightly coloured uniforms to motivate them and to make their presence and work more noticeable and easily monitored.

Twenty one (21) companies said the morale of the employees was high due to fair labour practices, 4 companies disagreed (Table 7.8). Fair labour practices in this context means whether employees are paid wages above minimum wage set by the government, and whether workers receive medical benefits, social security and other benefits, and protective clothing. On the issue of the morale of the employees being high due to fair labour practices, it was noticed that employees in some companies were paid wages above minimum wage; receive medical benefits, social security and loans with low interest or without interest. Generally, public sector waste workers and formal private sector workers are subjected to unhealthy working conditions and poor social security. Access to social and health care services is necessary to boost the morale of waste workers. The use of protective clothing and standard waste equipment may also help to alleviate the social stigmatization which is often associated with waste work. Standard waste equipment and protective clothing also reduce occupational health risks.

Detailed analysis of actual supervisory practices was conducted to further understand the companies' supervisory capability. The regression results show that supervisory capability did explain partly the variations in productivity. However, the results of actual practices of supervision of above and below average companies were mixed (Table 7.8), which presuppose that the responses of companies on technical personnel and management capability might be subjective. This triangulation analysis was done to check the validity of the companies' responses. The differences in monitoring and follow up on waste collection, and the managerial ability of top management, to some extent, explain why supervisory capacity influences productivity. The companies which were more capable of supervising operations perform a number of tasks. The capability of these companies to perform these tasks is necessary for increasing productivity. The roles of the supervisors are checking records in vehicle log books, monitoring and follow-up of waste collection. The supervisors also record the daily number of trips and tonnage of waste and also ensure that the waste pick up schedule and frequency of waste pick-up are strictly followed. Follow-ups and effective monitoring are done to ensure that the frequency of the waste pick-up is strictly followed.

There are log books for recording miles and the time used for the collection and disposal of the waste collected. The drivers have an expected number of containers they have to pick up or an expected number of houses they have to serve in a day. The drivers record the time of arrival and departure, and the time they pick up the first bins or communal containers. Other information such as time they get to the land fill and the number of trips they make in a day are recorded. Data of collection points picked up, time of visit and houses visited are recorded. The supervisors also go to the sites to check if the wastes are picked regularly and also they cross check their information with that of drivers. The supervisors check the log books and time sheets that the drivers fill out on daily bases or periodically. Analysis of data from the log books and time sheets is made, and the routes are either changed or maintained in order to reduce the hauling time and maximize vehicle productivity and reduce cost.

Table 7.8 Vehicle supervision capability of above and below average performers

Variables	Above Average Performers (N=6)		Below Average Performers (N=7)	
	Performed Task	Did not Performed Task	Performed Task	Did not Performed Task
Supervisors monitor and follow up on waste collection	6	0	7	0
Drivers have log books for records	4	2	5	2
Supervisors check log books	4	2	4	3
Supervisors analyse log books	3	3	4	3
Managers analyse log book	2	4	1	6
Supervisors record daily trips and tonnage of waste	5	1	4	3
Supervisor checks daily targets	5	1	6	1
Frequency of pick-up followed	4	2	4	3
Delayed scheduled pick-ups are picked within 24 hours	5	1	6	1
Use protective clothing	4	2	7	0
Waste spillover is collected	3	3	4	3

When there are delays due to traffic congestion then the waste is picked up within 24 hours but if the delay is due to truck breakdown then another truck will have to be re-assigned to pick up that waste and most of the time that will take more than 24 hours. The trucks are covered with nets during transportation to prevent the waste from spilling. The waste spillover at some sites is picked up and dropped into the new container. The drivers and collection crew put on protective clothing (clothes, boots and gloves) and also ensure that waste spillover unto the ground at collection point is collected.

Companies with less capability for supervision do no not have vehicle log books. Management sets daily output targets for each vehicle and this is what is expected of the drivers. There is no strict schedule of waste pick-ups. The most important thing is for the driver to make the expected trips. However, the follow-ups and monitoring are done to ensure that the frequency of the waste pick-up is followed, especially when there are complaints of uncollected waste from the residents.

Vehicle Maintenance Capability

Maintenance may be defined as the regular checks and repair work that is done on a vehicle or part of the vehicle regularly to keep the vehicle in good condition and to prevent total breakdowns. Vehicles used in waste collection need to be maintained (planned maintenance) to avoid breakdowns and extra cost and low productivity. Vehicle maintenance scheduling is a planned maintenance task which involves the preparation of maintenance schedules that are followed to ensure regular maintenance of the vehicles.

The results showed that 23 companies said vehicles maintenance schedules were prepared and followed (Table 7.9). Specific schedules of planned preventive maintenance and proper garaging facilities to undertake both routine maintenance and repair work can contribute substantially to increasing vehicle productivity. Maintenance of vehicles includes schedules and facilities owned for maintenance and repairing of collection vehicles. This is necessary because it leads to reduction in vehicles breakdown and increases service life of vehicles. Utilisation of vehicles depends on regular maintenance and response to faults and breakdowns of vehicles. For efficient operation, vehicle maintenance is required.

Table 7.9 Vehicle maintenance capacity

Indicators	Disagree	Agree
Vehicles maintenance		
Vehicle maintenance schedules are prepared and followed	2	23
In-House Maintenance		
Company has in-house maintenance workshop.	12	13

Source: Field Survey by Author

Again, results show that 13 companies had in-house mechanic workshops and mechanics with varied technical know-how for minor and major of vehicles maintenance depending on the fault that occurs. Others also resorted to external mechanic workshops as and when there was the need for both minor and major vehicle maintenance. In a situation where external workshops are used, if the repair of vehicles is not immediate, their productivity and utilisation in terms of waste collection will be highly affected. In addressing this situation some waste collection companies have established workshops where collection vehicles that are broken down are brought in for repair.

Companies which were more capable of vehicle maintenance perform a number of tasks (Table 7.10). Some companies have in-house mechanic workshops for vehicles maintenance. The managers prepare the vehicle maintenance schedule, and every vehicle goes through a routine check and servicing every month. For minor breakdowns vehicles are repaired within twenty-four hours. In the case of major repair works like engine failure or overhauling of engines it can take four days or weeks. The in-house workshop saves time and also prevents frequent breakdown of vehicles.

It was noticed that vehicle maintenance schedules are prepared and followed by some companies although breakdowns occurred despite planned maintenance schedules. Respondents pointed out that vehicle breakdown affects waste collection since it results in delays in the collection of waste. Vehicle breakdowns affect waste collection in that waste pick-ups are delayed and the frequency of breakdowns subsequently results in more delays. In some situations it becomes necessary to engage other vehicles and apart from the additional costs (overtime payments) that are incurred, the reliability of these vehicles poses challenges as well.

Table 7.10 Vehicle maintenance capability of above and below average performers

Variables	Above Average Performers (N=6)		Below Average Performers (N=7)	
	Performed task	Did not performed task	Performed task	Did not performed task
Maintenance schedules are prepared and followed	4	2	5	2
In-house workshop available	4	2	4	3
Vehicle maintenance elsewhere	5	1	4	3
Minor breakdowns are repaired within 24 hours	3	3	7	0
Major vehicles breakdown repaired within in 3 days	3	3	6	1
Routine/Planned maintenance	4	2	3	4

The influence of availability of in-house mechanic workshops on performance is not significant, since the results were mixed. Some of these companies do not have vehicle maintenance schedule, so the mechanics are called upon during breakdowns. For minor repairs, vehicles get back on the road within 24 hours of breakdown and if it is a major fault it can take about three days to a week. Since there is no maintenance schedule and only repair works are done, vehicle breakdown is frequent and this sometimes affects waste collection.

Road, Traffic and Vehicles Conditions

Road, traffic and vehicle conditions are also factors that influence productivity and utilisation of collection vehicles. The results presented in Table 7.11 indicated that ten (10) companies rated the access and surface conditions of roads used for waste collection as poor; nine (9) companies rated it as fair while six (6) companies rated the condition of roads used for waste collection as good. It was indicated that some access routes were bad and that breakdowns of collection vehicles were frequent for vehicles that were not designed for such routes. The bad roads had an adverse effect on the performance of the waste collection vehicles. Road conditions limit the maximum quantity of waste to be carried by each vehicle type.

Table 7.11 Road and traffic conditions of collection vehicles

Indicators	Response (N)
Road Condition	
Access and surface conditions of road used for waste collection	
Poor	10
Fair	9
Good	6
Traffic Condition	
The nature of traffic conditions and congestion on the roads along collection routes	
Low congestion	12
Moderate congestion	10
High congestion	3
Vehicles Condition	
Breakdowns of vehicles depends 'make' and 'age' and the higher the age of a vehicle the more breakdowns occur	
Disagree	5
Agree	15
Cannot tell	5

Source: Field Survey by Author

With regard to traffic conditions, twelve (12) companies reported that the traffic conditions along collection routes had low traffic congestion, 10 companies rated it as moderate traffic congestion and 3 companies rated it as high traffic congestion. On the issue of traffic situation in the cities and to manage the situation, respondents said that management of companies ensured that drivers report to work early enough to avoid high traffic congestion. Drivers were also abreast of traffic conditions at different times of the day and so planned their movement to avoid the situation.

Vehicles conditions were assessed in terms of whether frequency of breakdowns of vehicles depended on the 'make' and 'age' of the vehicles and whether the higher the age of a vehicle the more breakdowns would occur. The 'age' refers to how old a particular vehicle has been in service after manufacture and the 'make' refers to the manufacturer of the collection vehicle for example, M-benz and Renault. Fifteen (15) companies said the higher the age of a vehicle the more breakdowns occur and the breakdowns of vehicles depended on the 'make' and 'age' of the vehicles, whereas five (5) disagreed and 5 companies were undecided. They could not tell whether breakdowns of vehicles depend on the 'make' and 'age' of the vehicles.

The respondents said the breakdowns of vehicles influenced the number of trips expected in a day and that would ultimately affect the productivity of such vehicles in terms of waste collected in tonnes per day per vehicle. A breakdown that needed spare parts for replacement would greatly reduced vehicles productivity if the spare parts were not immediately available.

An interview with the drivers of the waste collection vehicles revealed that several hours are normally wasted as a result of high traffic congestion along some of the roads used by the collection vehicles. Collection vehicles usually do not meet the number of trips they expect in a day due to traffic condition in most cases, hence low vehicle productivity and utilization of some of the companies across the cities.

7.4   Conclusions

In this chapter, the effects of company capacity (scale of operation and capabilities) on performance were analysed. The indicators used to measure companies' capabilities were solid waste operations planning, supervision and maintenance. The market share and number of solid waste collection vehicles were used as proxy to measure the scale of operation. The research questions explored in this chapter are: Does scale of operation of a company in solid waste collection influence performance? Do private sector companies have the capabilities and skills for solid waste operations planning, supervision and maintenance?

The analysis has provided findings that raise questions and theoretical implications. The results show that there was increasing returns to scale up to 7 vehicles and thereafter decreasing returns to scale. This suggests that the optimum scale of operation is 7 vehicles, for the purpose of economics of scale. There was diminishing return to increasing scale of operations beyond the optimum scale. There was decreasing return to scale beyond 7 vehicles. This study suggests that the optimum scale of operation is 7 vehicles for the purpose of economics of scale. The scale inefficient companies had low productivity due to inefficient use of input resources. To maximise productivity the companies have to plan their collection routes for the drivers and do not have to allow the drivers to use their discretion as to which route to take. It can be concluded that the size of a company (market share) significantly explained, partly, the variation in productivity.

It can be concluded that high scale of operation of a company and supervisory capability of a company lead to high productivity. The analysis show that increasing scale of operation and capability for supervision and maintenance scheduling lead to increasing high productivity The findings confirm some aspects of the hypothesis that scale of operation of private companies and the capabilities for supervision influence productivity. The size of company (market share) significantly explained partly the variation in productivity. The capability of companies above and below average for operation planning and maintenance were mixed.

The theoretical implication is that adequate private sector capacity for service provision ensures that resources are used more efficiently and also stimulates innovations to take competitive advantage to achieve high productivity and provide better service quality. But the results suggested otherwise, since productivity was high for some companies but the service quality of these companies was not better. In theory, a well performing company achieves high productivity and provides better service quality. The inefficient scale of operation and inefficient operations management (supervision and maintenance) explain the inefficiencies of the companies across the cities. It was hypothesized that scale of operation of private companies and the capabilities for solid waste collection planning, supervision and maintenance influence private sector performance (productivity and service quality). Besides the influence of company capacity (scale of operation and capabilities) on performance, regulation is another factor that influences performance of private sector in solid waste collection. Chapter 8 will explore the influence of regulation (contracts, market concentration, service costs and charges) on private sector performance.

8   Regulation of Private Sector Involvement in Solid Waste Collection

8.1   Introduction

The analysis of productivity and quality of service of the companies was presented in chapter 6, and the influence of capacity on performance of companies was presented in chapter 7. This chapter presents the results of a survey of 25 companies and key informant interviews. This chapter focuses on the contracts, regulatory mechanisms and the regulatory capacity of the local government, and analyses the influence of regulation on the performance of the companies. These regulatory mechanisms are the access to information, regulation of user charges, and regulation of service quality. The research question this chapter seeks to explore is: How do regulation mechanisms and the extent of adherence to contractual rights and obligations influence private sector performance? It is hypothesised that adherence to contract obligations and regulatory rules provide incentives for cost recovery and better service quality.

The chapter is organised into six sections. Section 2 examines the scope of contracts and regulation by contract. Section 3 examines the regulation mechanisms and explores the influence of regulation mechanisms on private sector performance. Section 4 explores the influence of adherence to contract obligations on private sector performance, while section 5 examines regulation control and strategies. The last section draws conclusion from the analyses of the regulatory mechanisms and the influence of regulation on the performance of companies.

8.2 Regulation by Contract and Competitive Bidding

Regulation of private sector involvement has an element of regulation 'by contract' (Ballance and Taylor, 2005; Franceys and Gerlash, 2008). The regulation revolves around the degree to which regulatory decisions making meets the interest of the public and the regulated business, and the extent to which regulation of prices and service levels are specified in the contract or left to some form of extra-contractual resolution. This form of regulation is the 'public interest' regulation which seeks to protect the public interest. Solid waste collection by private sector requires institutions for contract preparation, monitoring and enforcement as well as mechanisms for dealing with the contract incompleteness that arises from the contract. A key role in ensuring that a contract works effectively is performance monitoring, which ensures that the private operator is meeting its obligations.

The model contract prepared by the ministry of Local Government (1998) in Ghana allows Local Governments (LG) to award solid waste contracts transparently through competitive bidding. In this study, the contractual arrangements between the local governments (hereafter called Assemblies) and the private sectors involved were assessed by the scope of services under contracts, competitive bidding, contract duration and renewal, adherence to contractual obligations, the enforcement of by-laws as well as the regulatory mechanisms.

8.2.1 Scope of Services and Contract Documents of Solid Waste Collection

Contract documents

A contract signed between a company and a Metropolitan Assembly provides legal agreement to enable the company to work according to the terms of the contract and to secure bank loans. A model contract was prepared by Carl Bro International/PEM Consult (1998) for the ministry of Local Government and it was mandated that the Assemblies should award solid waste collection contracts transparently through competitive bidding. The model contract stipulates that the contractor shall not have the right to terminate the agreement, unless the LG has not been able to make timely and regular payments and the contractor has not been able to meet its cash flow requirement for personnel and consumables. The clauses in the model contract for solid waste collection agreement are captured under the following headings:

o Scope of Services
o Obligations of the Service Provider
o Obligations of the Assembly
o Tariffs and Fees
o Monitoring
o Sanctions
o Period of Agreement and Renewal
o Force Majeure
o Governing Laws
o Settlement of Disputes
o Termination of Agreement
o Sign and seal agreement
o Annex to contract document (Designated collection areas, minimum staff and equipment requirements, specification of key personnel, estimation of refuse weight and price escalation)

There is another model contract document for waste and sanitation as a schedule of the Procurement Act (2003) and its content and structure is general and different from the 1998 model contract. Kumasi WMD is still using the 1998 models which is specific for solid waste collection whereas Tema WMD is using the model contract for water and sanitation. There is no uniformity in the contract documents and some of them lack detailed scope of services to be provided. The WMDs are involved in contract document preparation and management of contract. The Procurement Units of Assemblies are responsible for all contracts at the Assembly. The role of the WMDs is to provide information (on areas to be served and scope of work) for the preparation of the contract by the Procurement Units (PU) or to be prepared by the WMD and submitted to the PU for the bidding depending on the capacity of the WMD staff.

Contract Scope of Services

From the survey conducted with the various private companies across the cities, it was revealed that the scope of service under the solid waste collection contract varied. Three scopes of services can be distinguished: these are communal collection only, house-to-house collection only and contract involving both communal and house-to-house collection. For the contract involving house-to-house collection of solid waste, the waste is collected at the street kerbside in front of premises and payment of user charges is made by the households with/without subsidy from the Assembly. The house-to-house collection is used for service delivery in medium to high income areas which are well planned with easy access roads. The collection zone for the house-to-house collection is classified according to the willingness and ability-to-pay for the service delivery. The residential classification is done by the Assemblies taking into consideration the housing type, population density, economic status and access road. With the house-to-house collection, the waste generator places the waste bins outside where the private waste company picks it up for disposal once or twice a week. The company collects the user charges directly from the service beneficiaries. The user charges collection is usually done on monthly basis and hence the private company cannot afford to underperform because the service beneficiaries will not be willing to pay for the poor services quality rendered.

With communal waste collection mode, a number of central collection containers are placed within communities in a zone and these containers are accessible to the residents in that particular zone. As regards the contract involving communal collection, communal containers (when full) are picked-up from sanitary sites within communities and payment for quantities of waste collected is made through subsidy from the Assembly with/without user charges paid by the households. The communal collection is provided in low and some medium income areas with high population density and poor access roads. Also the communal service is provided in areas where willingness and ability-to-pay for waste collection service is very low. This mode of collection can be an environmentally unsound practice, especially within communities where waste generating rates are high with low collection frequencies.

For the collection contract involving combined communal collection and house-to-house collection the scope of services are as explained above. Out of the 25 companies, 18 of the companies had collection contracts covering both house-to-house and communal collection. The companies with house-to-house collection only were seven.

The broader scope of services of the new contracts covering both communal collection and house-to-house collection in Kumasi comprise: (a) the collection of solid waste from house-to-house; (b) the collection of solid waste from central containers to be placed at sanitary sites acquired and specified by the Assembly; (c) the collection of solid waste from industries and (d) the collection of solid waste from commercial facilities. The waste collected from house-to-house, central containers, industries and commercial facilities is to be transported to transfer, treatment or disposal sites to be designated by the Assembly. The areas to be served by house-to-house collection and by central containers are designated as zones and are served exclusively by specific companies.

8.2.2   Competitive Bidding, Contract Duration and Renewal

Competitive Bidding
The competitive bidding process is the means by which the private waste companies are engaged in solid waste collection. The companies' bids are assessed based on their capacity, resources and technical expertise, and the company that meets these requirements is awarded the contract for a particular duration. After the duration of a contract has expired, the company is required to undergo a contract renewal process through competitive bidding. However, competitive bidding to renew contracts was not being pursued by the Assemblies for most contracts prior to 2007 because after the initial contracts had ended the Assemblies gave notice letters to companies to continue service. The results show that out of the 25 companies surveyed 13 of them had no contract but they were rendering service. The companies without contract in Accra signed six months contract in 2001 with the Assembly and it was intended to be renewed every six months subject to service performance but the contracts for the next term were not signed, after the first term expired, due to lack of political will and commitment to scheduled payments as stipulated in the contracts. The companies were given letters to continue service in their respective zones until 2009 when competitive bidding was conducted for all the 10 zones in Accra. The absence of competitive bidding has an adverse implication on service efficiency as the quality of service is likely to be greatly compromised. The contracts of non-performing companies could not be terminated when Assemblies owe the companies for more than 12 months service rendered.

There has been recent competitive bidding in Accra and Kumasi except Tema, Tamale and Takoradi. The competitive bidding process was to engage the services of these companies for another five-year term in the solid waste collection services. The results show that the private companies underwent competitive bidding for some new contracts and renewals, but this practice was for a PSI project and a recent development. The competitive bidding for new contracts (one for each zone) is a sign of commitment of the politician and the WMDs towards private sector involvement.

Political Influence and Regulatory Capture in Contract Award
The selection of companies through competitive bidding is sometimes done along political lines. The key informants admitted that there were political interferences sometimes in the award of contracts. The issue of secrecy of award of contract came up during the key informant interview. The contract was signed by the Metropolitan Chief Executive, who is government appointee, without the involvement of the staff of waste management department. This company had links with the government in power and had contracts with the Assemblies while other companies had no contracts but were rendering service. The signing of the contracts was to safeguard its existence after a change in government and to avoid the abrogation of such contracts. There was an element of regulatory capture in the contract signed. The payment arrangement between the company and the central government was such that the WMD of the Assembly was not involved in the contract apart from the political head (metropolitan chief executive). The payments for work done by this company were regular compared to the payments to the other companies. The payments were made without any certification by the WMD, a department responsible for monitoring companies. The Metropolitan Coordinating Director and the Metropolitan Chief Executive who signed the contract wrote letters to the central government for payment to the company without the involvement of staff of WMD who monitor the service. It came out that the staff of WMD had not seen the contract and could not monitor the contract. The WMD was not involved in the contracting so the monitoring and enforcement of the conditions and terms in the contract were a difficult task for the WMD. The workers of this company did not recognize the staff of WMD and were rude to them. This is a clear case of regulatory capture with rent-seeking behaviour of the politicians.

According to Awortwi (2003:181) a systematic analysis of each of the contract processes showed that award of contracts as of 2003 was largely based on agent's connections with people in higher authority than on any

process of competitive bidding. Giving opportunity for renewal of contracts for incumbents without open competitive tendering provides an ave-
nue for corruption and opportunism. Two companies (one in Accra and the
other in Kumasi) did not go through any competitive bidding in 1999 and
2000 due to the political support at the time (Awortwi, 2003:155,181). The
results reveal that contract renewal in the five cities in Ghana through
competitive bidding was not done prior to 2007 but the companies were
allowed to continue working even though their contracts had expired.

The results from the survey show that the Metropolitan Assemblies
have begun to give more weight to the private sector involvement by open-
ing competitive bidding process for contracts renewals. The companies
submitted proposals for competitive bidding process in 2007 in Kumasi
and in 2009 in Accra. Private companies involved in solid waste collection
service may have their contracts renewed based on the condition that they
have performed satisfactorily (in terms of regular collection) in the past
and may have to go through competitive bidding.

8.3 Regulatory Mechanisms and Adherence to Contract Obligations

This section examines the regulatory mechanisms and regulatory decision
making of the local governments (Assemblies). These regulatory mechanisms are the access to information from the companies, regulation of setting of user charges, and regulation of service quality of the private sector.
The research question this section seeks to explore is whether these regulatory mechanisms influence private sector performance. The framework of
'agency capture' of regulation and 'public interest' protection by regulation were used to analyse the regulatory decision making on the regulatory
mechanisms.

8.3.1 Regulation of Access to Information

Table 8.1 presents responses to the variables used to assess the Assemblies' access to information from the private companies. In order for the Assembly to effectively monitor the quality of service rendered by the waste companies, the Assembly must have access to information from the waste companies. This information includes vehicle tour schedule, performance targets, tonnage of waste disposed of and revenue from house-to-house collection and cost incurred. The municipal /metropolitan Assemblies role is to monitor and regulate the solid waste companies. The access to information from the companies is important as this determines the level to which the Assemblies carry out these regulation procedures. The results showed that twenty-one (21) out of twenty-five (25) companies said the performance targets pertaining to the regulations of the private waste companies are clearly spelt out in the contract signed. Also twenty-one (21) companies said the Assemblies are able to get information on tonnage of waste collected. The Assemblies keep records on the tonnage of waste disposed of since the companies are paid based on the records. The companies are required to submit monthly reports on waste collected and revenue mobilized from house-to-house collection.

Table 8.1 Companies' views of regulation of access to information

Indicators of access to information	Adequate	Inadequate
Clear and measurable performance targets information in the contracts.	21	4
Assembly's access to tonnage information from company	21	4
Assembly's access to revenue information from company	25	0

All the companies involved in house-to-house collection said they provide information to the Assemblies on house-to-house revenue from waste collected and disposed of. However, key informant interview with WMD staff show that information on revenue is not properly reported to the Assemblies, and this is contrary to what the companies said. Solid waste collection cost and revenue information asymmetry was common in all the cities. Key informant interview showed that reports on claims and revenue collected are delayed at times and also the revenue information, most of the time, was far below the expected revenue to be generated. The companies gave lower revenue values because they are required to pay fee to the Assemblies on the house-to-house revenue generated.

The Waste Management Department Heads were of the view that even though clear and measurable performance targets are set in the contracts with the private companies, these targets are not fully achieved by the private companies due to inadequate physical capacity (trucks). For monitoring quality of service rendered by the waste companies, the Assembly must have access to all information from the waste companies. The monitoring information includes performance targets, number of trips made, tonnage of waste disposed of, revenue and cost incurred. The access to information by the Assemblies can also determine to which extent they carry out regulation procedures, but cost information asymmetry was common in the cities. Getting cost data from the companies was difficult and the Assemblies relied on their own cost data to determine user charges.

8.3.2   Regulation of Unit Price, User Charges and Cost Recovery

Regulation of Service charges and collection fee (unit price)

Table 8.2 presents responses on the views of the private companies on regulation of charges. Also twenty-one (21) companies were of the view that the unit prices and user fees for the communal and house-to-house respectively are set by the Assembly alone. Four companies responded that the fees for house-to-house collection without subsidy are set by the companies but they are approved by the Assemblies. These are normally those who are given the permission by the Assemblies to carry out house-to-house collection services but do not have contract with the Assembly.

Table 8.2 Companies' views on regulation of unit price and user charges

Indicators	Agree	Dis-agree
Tariff Setting		
Assembly set unit prices and user fees alone	21	4
Communal and House-to-House services subsidized by Assembly	25	0
Tariff Level		
Unit prices for communal collection are low and do not cover cost	18	0
Service fees set for house-to-house are low and do not recover cost	17	8

For contracts covering both communal and house-to-house collection with subsidies, the pricing mechanism is that the Assembly provides indicative monthly household tariff for house-to-house collection and unit price per emptying of each communal container in the bidding document. The companies in turn indicate in the bidding documents their proposed house-to-house tariff and unit price for communal collection. The tariffs and the unit prices are subject to changes after the award of contract using price escalation formula in the contract.

The communal collection was provided free for residents until 2008 when pay-as-you-throw user charges were introduced in Kumasi and monthly fixed charge was also introduced in Tema. The communal collection service cost is paid using subsidy from the Assembly and central government. The house-to-house collection is provided with or without subsidy from the Assembly. The house-to-house collection without subsidy from the Assemblies was rendered by the companies without contract prior to 2007 competitive bidding in Kumasi and 2008 bidding in Accra. These companies were permitted to provide service in particular delineated zones and user charges approved by the Assemblies. The user charge for the house-to-house collection without subsidy is fixed by the companies but approved by the Assemblies.

Service charges and collection fee (unit price) and cost recovery

Table 8.3 shows the user fee (service charge) by the beneficiaries with respect to the economic status of individuals, house-to-house and the communal collection. There were differences in the service charges for communal service collection and the house-to-house service within and across
the cities. For house-to-house, each household pays GH¢ 1.00 to 3.00 for
low income, GH¢3.00 to GHc6.00 for middle income and GH¢4.00 to
GHc12.00 for high income across the cities. Comparing the fee charged
across the five cities, it is obvious that the fee charged per bin for house–
to–house was high for people living in areas classified as high class. These
differences are attributed to the fact that individuals living in such areas
want services that meet their status in society of which environmental
cleanliness is paramount.

Table 8.3 Unit Price and Service Charges For Solid Waste Service in 2007

CITIES	HOUSE-TO-HOUSE CHARGE PER BIN (GH¢/bin/month)			COMMUNAL CHARGES (GH¢)	
	Low class	Medium class	High class	Price Per tonne	User charge
Accra	3.00	6.00	10.0-12.0	10.00	0.10-0.20 per headload
Kumasi	2.50-3.00	4.00	5.00	9.00	0.10-0.20 per headload
Tema	2.00	2.40	6.0-11.0	6.0-10.0	0.80 per month
Takoradi	1.50-2.00	3.00	4.00	9.00	0.05-0.10 per headload
Tamale	1.00-2.00	3.00	4.00-5.00	9.00	No user charging

US$1 = GH¢1.2

Under the house-to-house arrangements the companies collect user
charges from the users and subsidy from the Municipal Authorities for
some areas. The house-to-house service charges, ranging from GH¢ 1.00 to
GH¢12.00 per month per bin, are based on the level of income and vary
from city to city. The user charges collected from beneficiaries under the
house-to-house collection were approved by the Assemblies. The companies rendering the house-to-house collection collect these approved fees

from their customers on monthly basis or any other agreed time. The users were not sanctioned when they did not pay the user charges. Some users default for more than three months but the companies have no mandate to sanction defaulters.

However, tariffs should be set at levels that will not discourage the use of the services, especially where this would create health risks through negative externalities. Also house-to-house collection fees charged for low income class range from GHc 1.00 for Tamale to GHc 3.00 for Accra and this can be attributed to the differences in the living standards, ability and willingness to pay and also travelling distances to the final dump disposal sites.

According to the waste companies and the Assemblies, the price per tonne of communal waste collection ranged from six to ten Ghana Cedis (Gh¢6.00 to 10.00), but prior to 2008 the communal users in Accra, Kumasi, Tamale and Takoradi did not pay for dumping into the containers. Currently, the users of communal collection service in Kumasi pay user charge of GH¢ 0.05-0.20 per head load, since the service is rendered in low income suburbs and most people are low-income earners. The companies rendering the service in Kumasi were tasked by the Assembly to collect the user charges. The Local Government (Municipal Assembly) in Tema collects GH¢ 0.80 per month as user fee from residents receiving communal waste collection service.

The review of user charges was not regular so the companies faced cash flow problems due to the escalating foreign exchange rate and increasing fuel prices and inflation. The new user charges need to be approved by the Assembly and published in national gazette as by-law for it to be legally binding on residents, and this most of the time requires the political will on the part of the City Mayor and the Assemblymen. The political will to implement full cost recovery through charging of all households was lacking in all the cities prior to 2008.

Pay-as-you-throw as cost recovery mechanism
The users of communal collection system were not charged user fees prior to 2008 as all service costs were borne by Assemblies. The pay- as- you-throw for communal collection has been introduced in some parts of Accra,

Kumasi, Takoradi, and Tema with the exception of Tamale. The pay- as-you- throw in Kumasi was approved by the Assembly and by-law passed. The pay-as-you-throw in Accra was termed as 'illegal' by the head of Waste Management Department since there was no legal approval, whereas in Takoradi it was approved and the Assemblymen tasked to collect the revenue but they did not account for monies collected. The revenue from the pay-as-you-throw for solid waste collection in Accra and Takoradi did not go to the Assemblies or the companies as the case of Kumasi.

The pay-as-you throw is a practice in USA and parts of Europe and now a mechanism that promotes solid waste source reduction, recycling, and diversion from landfill (Skumatz, 2002). PAYT, strictly speaking, despite its challenges has proven to have potential for cost recovery in low income communities in a developing country. Under the PAYT user charging arrangements, communal system users pay 10-20 Ghana Pesewas per head load of solid waste dumped into communal container. This user charging arrangement has become necessary to help reduce cost of waste collection in the cities. This is intended to supplement the Assembly's subsidy for solid waste collection. The low income residents who contribute the large quantities of waste in the cities could pay through PAYT for improving solid waste collection.

8.3.3   Regulation of Quality of Service and Contract Monitoring

Regulation of service quality is intended to ensure good practices in the delivery of solid waste collection services, thus ensuring quality service delivery by the private waste companies. Regulation of quality of service in this study was assessed in terms of monitoring of companies compliance to regulations, monitoring of quantity of waste collected, and Assemblies adherence to the terms and conditions in the contracts signed with the companies. The results of the views of the private companies on regulation of service quality across the five cities are presented in Table 8.4. Twenty (20) respondents said monitoring of the companies compliance to regulations and standards set in the contracts was adequate. The monitoring of companies' quantity of waste collected and their output were viewed as adequate by 22 respondents. The Assemblies also have an obligation of supervising the private companies in carrying out their waste handling functions. Eleven (11) companies responded that the Assemblies inadequately carry out their supervisory role. This role is not effectively carried out and might be due to peculiar problem faced by the Assemblies. This problem is lack of resources (capacity and funds) to carry out these functions. Assemblies' adherence to the regulations in the contracts signed with the companies was viewed as inadequate by 13 of the companies. The household survey of residents' appreciation of service quality in the cities showed that the level of service quality is below expectation of the residents, and waste overflow was high for communal collection than house-to-house.

Table 8.4 Companies' views on regulation of quality of service

Regulation indicators	Adequate	Inadequate
Monitoring of companies service quality and compliance to regulations and standards set in the contracts.	20	5
Monitoring of quantity of waste collected and output of company.	22	3
Contract management and supervision of the private sector by the Assembly	14	11
Assembly's enforce sanctions in the contracts signed with the companies	12	13

The Assemblies are required to out-source services through contracts and also conduct regular monitoring of the quality of service rendered by the private service providers and sanction any non-compliance according to the regulations and terms of the contract. The private companies are required to submit monthly and annual reports to the Assembly as part of monitoring of the services provided by the private companies. However, there were lapses in the monitoring and the data needed for the monitoring. The results show that the Assemblies had limited capacity to regulate and monitor solid waste companies. Regular monitoring of quality of service was hampered because of inadequate capacity and resources of the Assemblies and the long delay in paying for the contracts. The Assemblies made payments to the companies for the services they provided but payments were usually delayed.

The Assemblies' role of public education on solid waste management issues and the enforcement of laws against indiscriminate dumping of solid waste were relaxed. The communal collection generates an environmentally and aesthetically unpleasant practice especially within localities where waste generating rates are high with low collection frequencies. Most often than not, the waste containers overflow owing to low collection frequencies and hence residents dispose of their waste onto the ground until the container is emptied and this encouraging free riding. Also some of the containers are lifted without cleaning the site as specified in the contract leading to heaps of refuse at the container sites. These sites breed mosquitoes, produce bad odour and hence threaten the health of residents staying around the site.

The private companies who are into the house-to-house collection deliver better services than the companies doing the communal collection. This is because they operate mainly in medium to high class residential areas where willingness and ability to pay is high, and therefore cost recovery is likely to be realised from beneficiaries who desire and demand quality service delivery and also value for their money.

Service standards are not well defined for the solid waste service delivery even though few are specified in the contracts. The covering of containers during transporting to disposal site is monitored. Those who receive house-to-house service use standard bins and different receptacles (eg. bags, paper boxes etc). The use of unapproved receptacles by households without fence wall around the house encourages littering of waste by dogs and other animals.

Service delivery monitoring is important for ensuring service quality and compliance with conditions and terms in the contract. The monitoring of service quality in Accra is done by the WMD through: (1) daily inspection of container sites and containers lifted, (2) residents complaints about areas not served as scheduled also helps with the monitoring, (3) twice weekly inspection of house-to-house service areas during the scheduled days for solid waste collection, (4) companies which do not comply with the frequency of collection and terms of the service delivery are served with default notice.

In Tema, contract monitoring by WMD involves field inspection of solid waste collection and following up residents' complaints of un-collected waste by two supervisors working with the WMDs. The fre-quency of waste collection is twice a week but some companies do once a week collection, due to delay in payment. In some cases when payment was delayed and the companies were in financial difficulty in buying fuel for their operation, fuel was given to all the companies by the Assembly and the money deducted from payment of arrears. The delay in payment affects service quality rendered by the companies because it is difficult to sanction them for non-compliance to terms and conditions in the contract. The enforcement of service standards has not been well addressed by the WMD.

In Takoradi, the monitoring of the contracts and service delivery is car-ried out by the Waste Management Department of the Assembly and the Waste Officers at the Sub-metros. The monitoring is done through visual inspection of waste collection and waste records at final disposal site, complaints by customers, meetings with the companies, monthly reports submitted by the companies, estimation of volume of waste collected using the vehicles capacities and number of trips.

In Tamale, the Environmental Health Department (EHD) of the As-sembly is tasked with the monitoring of collection service. The field in-spection of the solid waste collection situation in the zones is conducted by the EHD. The collection is also monitored through landfill records on number of trips and submission of reports by the private companies.

The breakdown of companies' vehicles affects their quality of service. WMDs are required to intervene when such events occur, but they also lack the physical logistics (trucks) to deliver. The capacity of some com-panies is inadequate. When collection trucks broke down due to old age, and no arrangements are made with the WMD or other private company for waste collection, then waste spillover and accumulation on the ground is bound to happen. The ground evacuation of solid waste was usually done by the WMD and cost surcharged the company operating in the area. Even though the WMD may recover the cost of ground evacuation from the company operating in the zone, the waste spillover and irregular collection of solid waste did not encourage user charging and willingness to pay by the residents.

The WMD is responsible for monitoring the private sector involved in the collection and disposal of solid waste. Monitoring of solid waste collection include: daily inspection of service areas, inspection of container sites, serving default notice to firms, serving warning letters, and terminating contracts for poor performance. Most of the functions have not yet been decentralised to the sub-metro level. For example, staff at the sub-metro level indicated that when the private firms are not delivering, the sub-metros can only relay the information to the WMD. The sub-metros staff supervise the private firms and report to the WMD. The monitoring of service quality is also done through Assemblymen and residents' complaints and assessment of the private companies. The monitoring of service quality is an issue of concern in a contractual arrangement. The value for money in solid waste collection service requires transparency in contracting and service quality monitoring.

Contract performance monitoring requires that service standards and indicators of outputs and outcomes are specified in the contract, and these indicators are used for service quality monitoring (Cointreau and Coad, 2000; Lusthaus et al., 2000). In the case of Ghana, most of the contracts do not specify the indicators and the outputs to be measured. The items listed in the contracts are input resources (workshop, staff, and vehicle requirements) for the delivery of the service, and those that concern output are changes in registers of those receiving service, summaries of waste collection carried out, summaries of complaints received and remedial actions taken, changes in staffing and summaries of vehicle performance and down time. The performance monitoring indicators are not well defined in the contracts. For example, in one of the recent contract, the clauses on monitoring are that:

o *Performance monitoring.* The services of the service provider under this agreement shall be subject to performance monitoring by a monitor designated by the assembly. The assembly may assign the responsibility for monitoring the performance of the services to the assembly's own staff, or may separately assign responsibility for monitoring to a private firm. The service provider shall co-operate fully with the monitor.

o *Input records.* The service provider shall allow the assembly, including the designated monitor and the assembly's staff, to have access at all times to inspect work being carried out under this agreement, to inspect all records and documents maintained by the service provider regarding work performed under this agreement and to inspect the service provider's vehicles, plant, stocks of spare parts and workshop facilities.

o *Monthly reports and output records.* The service provider shall submit to the assembly monthly reports in a format acceptable to the assembly, including changes in the registers of those receiving service, summaries of waste collections carried out, summaries of complaints received and remedial actions taken, changes in staffing and summaries of vehicle performance and down time.

o *Monthly meetings.* The service provider shall attend, at the request of the assembly, monthly meetings where operational and other issues of mutual interest may be discussed.

8.3.4 Assembly's Adherence to Contract Obligations

The Municipal/Metropolitan Assemblies have the obligations to promulgate by-laws to assist the private companies in their service, conduct public education regularly, pay companies according to the specifications in the contract. All these are required of the Assemblies to comply with as stipulated in the contracts. The companies were interviewed to assess the Assemblies' adherence to contract obligations and the results are summarised in the Table 8.5.

Table 8.5 Companies views on Assemblies adherence to obligations across the cities

Indicators	Agree	Disagree
Provides designated and accessible disposal site	25	0
Provides designated and accessible container/sanitary sites	25	0
Promulgates by-laws to assist companies in their service.	22	3
Conducts public education regularly.	14	11
Pays companies within 45 working days after date of claim.	0	21
Pays interest charges if payment delayed beyond 3 months.	0	18

The respondents said that the Assemblies promulgate by-laws to assist companies in their service. The five (5) cities have by-laws to assist the companies in the delivery of their services. Local government (LG) Act 1993 (Act 462) mandate the Assemblies to promulgate by-laws to regulate solid waste and sanitation service. These by-laws were written out of the national laws on sanitation and are to be adhered to by the residents to ensure good sanitation practices. Each flout of the by-law by residents attracts a sanction from the Assembly. However, twenty-two respondents said there are by-laws but these by-laws do not promote effective solid waste collection, in view of the rampant indiscriminate dumping of solid waste.

It is the obligation of the Assembly to conduct public education regularly to educate the populace on sound environmental practices but from the responses gathered from the interviewees only 14 out of the 25 agreed that the Assemblies conduct public education but not regularly. This is clearly evident in the people's attitude towards environmental cleanliness. In as much as the private companies do their best to keep the cities clean, the attitude of the people towards good sanitation practices leaves much to be desired. This is attributed to the lack of regular education on good sanitation by the Assemblies and the companies.

The views from the companies revealed that part of the contract is partly not fulfilled in that, 21 companies said the payments are delayed beyond the stipulated time and this accounted for 84% of the respondents. As to the payment of interest charges if payment is delayed beyond the 3 months, 18 companies said it is not done and this represents 72% of the respondents but the private companies do not take any legal action against

the Assemblies. The Assemblies are also obliged to pay the private contractors within 45 working days after date of claim and if payment delays
beyond three months, they are to pay the companies interest charges as
clearly stipulated in the contract. The clause on payment of interest has
been repealed from the new contracts in 2008 and this is a clear case of
regulatory capture by the Assemblies. In this case of 'agency capture regulation' the chief executives of the Assemblies control the solid waste collection services and payments to the interest of politicians they represent. Interview with key informants revealed that one company received regular payments and this company had political connections with government officials
at the time. The regular payment to this company could also be attributed to
'regulatory capture' by interest groups in government with rent-seeking and
collusion politicians.

The main problems faced with the communal collection by all the private waste companies interviewed were low cost recovery, delay in payment
by the Assemblies leading to occasional interruptions of service delivery.
The monthly average solid waste collection cost for Kumasi in 2008 was
GH¢344800 for solid waste quantity of 42100 tonnes. The average monthly
total cost of GH¢589,200 arising from the waste collection and disposal of
36000 tonnes in 2007 was a news in the media (Kite News, 2008). The collection monthly cost alone at GH¢ 9.00 per tonne was GH¢ 324,000. In Accra the Principal Environmental Health Technologist at the AMA said the
Assembly owed various waste companies over 7.5 million Ghana Cedis as
of 2008. In 2010, out of 24 million Ghana Cedis (17 million US dollars)
owed, the Accra Assembly paid 10 million to the companies. This irregular
payments for services rendered have made it extremely difficult for the service providers to deliver quality services. The private companies delivering
waste collection services are paid based on the number of trips recorded at
the final disposal site. A form is signed by the WMD officer at the dump
site. This is to serve as an authentication for payment by the WMD for services delivered by the private companies.

The problem of cash flows, delay in payment and the erosion of the
revenue due to non-payment of interest affects the performance and quality
of service rendered by these private companies. This resulted in reduced

frequency of solid waste collection (once collection instead of twice for house-to-house collection) in an attempt to reduce cost of collection. This in turn results in waste overflow at communal container sites which was evident from the household survey. Twenty of the companies are not able to procure new trucks for the collection and this affects the service quality as a result of breakdown of over aged vehicles. The delay in payment of subsidy from the Assemblies is a major setback which is a source of worry to the private companies. According to the private companies, they sometimes have to turn to the banks for loans to keep them in business and they pay these banks when they are paid by the Assemblies. Most of these companies have vehicles which are not road worthy. The risk in investment in new trucks has compelled most companies to use over-aged vehicles. Even though the solid waste business is viable, the delay in payment and the unsigned contracts make the business unattractive for banks to grant loan facility for procurement of new trucks.

The house-to-house collection has a better financial arrangement than the communal collection. However, with the introduction of pay-as-you-throw in Kumasi, the extension of pay-as-you-throw to the other cities will reduce the subsidies and the associated delays in paying the companies. The collection of user charges from households is an area of concern when it comes to revenue generation for solid waste collection. According to the companies, they have been successful in both high and middle income areas, managing to recover about 90% of the fees (the only problem being some delays in payment).

The bottlenecks identified in the financial arrangement are that the Assemblies alone set the user charges without consulting the private companies and there was delay in upward review of user charges and collection fees, as well as delay in payment for solid waste collected. According to the companies, the tariff level is inadequate for cost recovery. There was weak mechanism in place to deal with residents who refuse to pay for the waste collection services rendered. The Assembly delays so much with the payment of monies due these private companies, and this in effect affects service quality. It can be concluded that Assemblies' non-adherence to contract obligations did influence service quality and productivity of companies, since all the Assemblies defaulted in subsidy payment and interest payment. The differences in performance (productivity) of the companies were due to the differences in the internal capacity of the various companies and the regulation of the private sector. The WMDs and the Assemblies are faced with challenges of user fee charging and low cost recovery. The low cost recovery through user charges for solid waste collection and the inadequate funding affect the service quality and performance of the companies. In Accra for example, sixty percent of the Assembly's revenue is spent on solid waste collected, but this is still not enough for solid waste collection. As the head of one WMD said:

> We are faced with lack of funds and inadequate revenue to pay for the services. There was un-willingness to pay for house-to-house service due to poor service by some companies. The users are willing to pay provided the service quality is good.

There was concern of apathy of the people concerning payment of user charges. Some of those who were paying under the house-to-house arrangement had stopped paying. This was attributed to the poor service quality, lack of enforcement of by-law, and inadequate public education. The lack of enforcement of by-law and inadequate public education were due to the weak coordination and capacity of the WMD and the Environmental Health Department of the Assembly. The enforcement of sanitation standards and payment of user charges require public education and awareness. There is arrangement for prosecuting the residents who default in paying user charges. The procedure is that a list of defaulters is usually submitted by the company through the WMD to the Assembly and the Environmental Health Department of the Assembly processes the defaulters for prosecution in the law court.

Analysis shows that solid waste collection is financially viable, since the private providers made some profit after tax, but the profit may be eroded by the delay in payment and the delay in the upward review of user charges due to the weak regulation by the Assemblies. The differences in the levels of the user fees and the mix of house-to-house service and com-

munal service may indirectly affect the quality of service. The difference
in the charges may lead to inadequate private finance which reflects in the
use of over aged vehicles (above 10 years). The use of over aged vehicles
results in poor coverage and low quality services due to frequent vehicle
breakdown. Some companies rely on loans but the payment by Assemblies
for waste collected delay most of the time for more than a year. The key
issue with lack of investment is new trucks were the unsigned contracts,
lack of government guarantee for bank loans and delay in payment.

8.3.5   Enforcement of Legislation and Sanctions

Enforcement of by-laws and contract rules is a mechanism that the Mu-
nicipal/Metropolitan Assemblies apply to control the management of solid
waste collection services. Assemblies' obligation to enforce by-laws was
assessed from the point of view of the companies. Table 8.6 presents the
views of the companies interviewed on enforcement of by-laws and conditions in the contracts. The views of the companies across cities did vary as
thirteen (13) agreed to the fact that the by-laws were enforced, nineteen
(19) also agreed that the fines for defaulting the by-laws were punitive
enough, fourteen (14) agreed that monitoring of compliance was done effectively and finally twenty (20) of the companies interviewed said the environmental health standards and sanitation were strictly observed and enforced. From the view point of the companies, the by-law enforcement role
of the Assemblies was being pursued; however, the household survey
shows not much is being done to enforce by-laws. The companies were not
confident to criticise the local government.

Table 8.6 Companies views on enforcement of legislation

Indicators	Agree	Dis-agree	Cannot tell
The by-laws of the Assembly and national laws on environmental sanitation are enforced.	13	8	4
The fine for defaulting by-laws is punitive enough.	19	2	4
The monitoring of compliance to solid waste service standards set in the contracts is done effectively.	14	6	5
The environmental health standards and sanitary regulations are strictly observed and enforced.	20	2	3

The Assemblies are not effective in sanctioning offenders or residents because of the degree to which these by-laws are flouted. Also the judicial process in Ghana is perceived to be very slow to deal with offenders. The lack of sanitation courts in the cities seems to be another factor which hinders the enforcement of solid waste and sanitation regulation. There are by-laws on sanitation to sanction offenders but in practice these are not strictly enforced. In Ghanaian cities, many citizens are not law abiding and show lax attitude toward sound environmental practices. This can be attributed to the fact that there are inadequate waste containers, low frequency of waste collection especially with the communal collection, long distance to central containers and lawlessness in the fact that citizens are not afraid that their actions will lead to sanctions by by-law enforcement authorities because these by-laws are, in practice, not strictly enforced.

There are sanctions in the contract documents which are applied to the service providers when they fail to comply with any of the obligations set out in the contracts. The fines for non-compliance differ from city to city and the sanctions are subject to review and may be changed by the Assembly after the service provider has been informed of such changes. The fine ranges from GH¢ 5.00 - GH¢ 20.00 across the cities. For example in Accra, a sanction of ten cedis (GH¢10.00) per day was imposed on companies which defaulted in picking up waste, but it was difficult to enforce this sanction due to the delay in payment of service contracts. The Assemblies have not been able to terminate non-performing contracts/services on the basis of non-performance with regard to enforcement of the terms in the contract. They are unable to apply the stipulated sanctions in the contract and also sometimes fail to adhere to their obligations in the contract. A bottleneck identified with the contractual arrangements is that after a term of contract has expired the company still remain in business. The obligations of the Assemblies and companies in the contracts are in practice not strictly adhered to by the Assemblies.

There was evidence of zonal monopoly violation (open competition for house-to-house service customers) across the cities but no action was taken by the Assemblies. Some companies were entering the zones of other companies to distribute bins to households for house-to-house collection in Tamale, Accra and Takoradi but the Municipal Authorities did not call the companies to order, due to the support of the Metropolitan Chief Executives and this is a clear indication of support from politicians. The companies who had the contracts or permits to work in these zones complained but there was no attempt to prevent the encroachment into their zones. There was no such violation of zonal monopoly in Kumasi due to the market discipline and the effort of the Kumasi WMD.

According to key informant interview, the following weaknesses in enforcement were identified: (1) weak enforcement of indiscriminate dumping by the environmental health department encourages residents to avoid user charges, (2) negative attitude of the people is a real challenge to the waste companies and the WMD, some people do not put the solid waste into the containers provided but put them on the ground, and (3) low public awareness about solid waste collection service, environmental sanitation, and user charging. This is partly due to the high cost of media charges for jingles and awareness programmes. One company director was of the view that:

> Some households refuse to comply with the by-law but it is not enforced. The negligence of the enforcement makes solid waste collection difficult. Those households who do not pay corrupt the workers and receive service first and better service. He was of the view that environ-

mental health officers could be seconded to work with the companies and paid by the companies to prosecute defaulters. He questioned if something cannot be done to get the Authority to pass a law for solid waste management.

These views were expressed by a number of directors from the key informant interviews. Another director said:

> The environmental health officer/health inspectors need motivation from the company before they go for inspection. The control of enforcement could be delegated to the private companies. The company will pay the environmental health officer who is seconded to the company to enforce by-law in the collection zone of the company. To facilitate the work of the environmental health officer, the waste of households who do not pay will not be picked, and the environmental health officer will go round and arrest those whose waste is not picked on the collection days. Besides, the Assembly has to promulgate a bylaw to eliminate all illegal solid waste collectors operating in the company zone to ensure total solid waste collection and cost recovery.

The attitude of the residents towards solid waste needs to be changed through behavioural change programmes. They dump waste by the road side and places they are not supposed to dump, some do not bring the waste to the road side on the collection days. According to the head of WMD in Takoradi, public education through community forum and radio jingles was organized in the past but the response was not encouraging. The un-cooperative attitudes of the residents and dumping on the ground when container may not be full lead to poor service quality.

This study is consistent with the finding of Awortwi that there is lack of sanctions. According to Awortwi (2003:192) case studies of 3 cities showed that many of the contracts had credible threats of sanctions that required sanction, but non-complying firms were not penalised. The theoretical implication is that for a principal (Assembly) to be able to sanction an agent (private company) it first has to fulfil its own obligation.

8.3.6   Company's Contractual Obligations

The details of all activities that should be carried out by the private waste companies are specified in the contracts. These include achievement of daily collection target, the use of professional and technical personnel in service delivery and ensuring that the container sites are free of litter. Also the company is to ensure that the collection crew put on protective clothing at all times during the discharge of service, cover all waste containers during the course of transportation. Table 8.7 presents the views of the companies across the cities on the assessment of their adherence to these obligations under the contract.

Table 8.7 Companies views on their adherence to obligations specified in the contracts

Indicators	Agree	Disagree
Company achieves daily collection target in the contract.	19	6
Company's use professional and technical personnel required.	25	0
Company cover waste containers during transporting	25	0
Company collection crew use protective clothing	23	2
Company keeps container site free of litters and clean.	24	1

Nineteen (19) out of the twenty-five companies interviewed were of the view that they were able to achieve their daily targets. This accounted for 76% of the respondents and six (6) representing 24% of the total number of respondents said they were sometimes unable to meet the daily target due to vehicle breakdown and lack of funds for regular maintenance. All the companies used technical and professional personnel in the discharge of their duties. However, the qualifications of the supervisors and managers of some of the companies were low. This suggests why some companies had low productivities.

As to whether the collection crew uses protective clothing during the collection of waste, 23 out of the 25 companies said the collection crew used the prescribed clothes. The other two in Kumasi and Tema, sometimes do not use the prescribed protective clothing. The workers thought that they had become used to the waste collection without protective clothing and therefore immune to any form of infection that may result from waste handling. The private companies are obliged to keep the container sites free of litter and clean especially with the communal collection. Twenty-two (22) companies keep their zones of operation free of litter whiles the remaining three are not sometimes able to keep their zones of operation free of litter. The latter can be attributed to the fact that large heaps of solid waste remain at the container site after solid waste has been picked up and this requires extra cost and effort to keep the site clean. They have to employ pay loader and tipper truck to evacuate the waste from the ground. The companies are required under the contract to repair and maintain communal collection containers; however, interview with key informants from the WMDs showed that the repairs of communal containers by companies were not done.

8.4   Regulatory Control and Strategies

8.4.1   Regulatory Capacity and Autonomy of Assemblies

The regulatory capacity and autonomy of the Assemblies to regulate the private sector activities was assessed using four indicators in Table 8.8. Most of the companies perceived the regulatory capacity of the Assembles to be adequate, whereas financial autonomy was perceived to be low. The main challenge facing the WMDs and the Assemblies is financial constraints resulting in irregular payment to private companies and inadequate funds for daily running of vehicles for service being rendered. Urbanization and influx of people daily into the cities require enough financial resources for solid waste collection at households, the markets and transport

stations, and from low income residential areas. The WMDs are using obsolete and inadequate vehicles for WMD in-house collection. The Assemblies did not have the capacity to collect 20% of the waste as stated in the
environmental sanitation policy. The lack of logistics for monitoring affects the work of the WMDs.

Table 8.8 Regulatory Capacity and Autonomy of Assemblies

Regulatory capacity and autonomy indicators	Not ade-quate	Ade-quate
Regulatory Capacity of Assemblies		
Accountability of the Assembly to EPA and head ministry	2	17
Independence of the Assembly in regulating the company's ser-vices	7	17
Resources and capacity of Assembly to monitor service quality and non compliance of service delivery,	7	16
Frequency of monitoring companies' compliance to service quality and standards	4	12
Financial Autonomy of Assemblies		
Autonomy to set level of user charges	5	8
Autonomy to set level of price per tonne for service contract	3	11
Autonomy of the Assembly (level of budget for solid waste)	11	4
Autonomy to implement national sanitation policy on cost recovery	2	10
Managerial Autonomy of Assemblies		
Autonomy of the Assembly to terminate contract of non-performing company	3	11
Autonomy to enforce by-laws and prosecute free riders	2	10

The management of the open disposal sites is a concern to the companies.
The absence of sanitary landfills in Accra, Tema and Takoradi makes
management of the disposal sites and access more difficult especially during rainy season. The disposal sites are difficult to drive through and companies dump waste in the middle of the sites due to the muddy nature of
the sites arising from the rainfall. The trucks get stuck in the mud most often and they had to be pulled out by other trucks. For example in Ta-

koradi, the two dozers for pushing and compacting the waste and making access road in the dump site for trucks, did not work at times due to delay of payment for work done, breakdowns and at times delay in supply of fuel. The non-operation of the dozers at the dump site affected the solid waste collection. The trucks could not pass through the accumulated waste at the dump site so drivers had to stop lifting of containers resulting in accumulation of waste within the communities in the city. This raised concern by the people, the media and local FM radio stations. The dumping sites need to be managed during the raining season to ensure continuing lifting of containers and solid waste collection from houses.

There is the need for more qualified personnel for the solid waste management. The Assemblies, with the exception of Kumasi, are not well staffed with the right personnel for the work. The ability and competence to collect data, analyse and take evidence-based decisions require high level training, possibly master's degree. The Kumasi WMD has staff with the highest qualification. The director, his deputy and two other personnel involved in SWM in Kumasi are MSc holders. Their counterparts in the other four cities are environmental health officers (few with first degree but mostly diploma holders). No effort has been made in the other cities to employ the right calibre of personnel for SWM. An attempt was made in Accra during GTZ project in the 1980s to train and retain qualified personnel for WMD in the other cities but it did not succeed. In Kumasi, effort has been made to motivate and encourage staff to develop capacity up to MSc degree level through past projects. The competence of the staff of WMD is important to be able to plan programmes and execute them. For example, the pay-as-you-throw approach for charging communal collection users has been successfully started in Kumasi due to the capacity of staff at the WMD.

The capacity of the Assemblies to facilitate, regulate and monitor solid waste collection is hampered by a number of factors. These factors include inadequate staff capacity, lack of funds, inadequate and over aged collection vehicles, weak institutional structure and capacity (legal, policy, regulation), and lack of political will. According to one of the WMD directors:

The main constraints of solid waste collection are weak capacity of Waste Management Department of the Assemblies and lack of political will. The experts in the SWM advise politicians in the Assembly on what to do, but because of political expediency the politicians do what they want if the experts are not able to convince them on the benefits of solid waste management interventions. The political will to regulate user charges and enforce by-laws is lacking. The weak regulation of user charging and weak enforcement of by-laws has affected the service quality of most of the companies in all cities.

Similar views on the capacity for solid waste management were expressed by another director of Waste Management Department. According to him:

There is lack of political will to take certain decisions on solid waste management. For example, service charges and collection fees were increased and decreased again in the 1990s. Again, the collection fee fixed in 2002 has not been reviewed upward till date, and cost recovery is low. The attitude of Assembly officials toward work does not help, they lack commitment to work. In addition, the hidden cost of arrest of defaulters of service charges payment makes it difficult to collect service charges. Another problem is the inadequate capacity of the Waste Management Department and the companies. The private sector has not developed, to integrate waste recycle and hazardous waste management into solid waste collection. He questioned if the Assembly can guarantee to buy 50% of compost produced by an investor for city beautification.

The weak capacity of the Waste Management Departments and political will contribute to the problems of solid waste management in Ghana. The capacity problems facing solid waste collection according one company director are that:

The zoning by the Assembly does not combine high and low income areas. The zoning using the sub-metros boundaries has a problem of assigning a high income area to a company instead of proper demarcation to include high and low income groups. The collection fees for communal collection and service charges for house-to-house collection are fixed arbitrary without any basis and are not adequate. These fees and charges are fixed and are imposed on the companies by the Assembly. For the poor areas which genuinely cannot pay fixed service charges, the Assembly has to pay for them through subsidy. The fee and charges are low such that the companies cannot purchase new collection vehicles and hire more quality human resources. The old vehicles break down often which makes the collection unreliable. There are also the problems of lack of user involvement in the service delivery, lack of political will, and policy issues.

These capacity and autonomy problems of the Assemblies which affect solid waste collection were reiterated by the Environmental Service Providers Association (ESPA). According to ESPA:

The companies do not have the financial capacity to purchase new collection vehicles due to inadequate collection fees and service charges to recover cost. Again, the drain desilting and street sweeping should be included in the solid waste collection contract. The association has some challenges due to the politics in the waste sector. For example, the expired 6-months contract in 2001 was not renewed even though the companies were asked to continue with service; one company was allowed to distribute bins for house-to-house collection in zones of other companies; two companies were allowed to place skip containers in zones of other companies for market and communal waste collection; four companies among ESPA received part of their money owned them by the Assembly before the 2008 competitive bidding to purchase new collection trucks, so that these companies would have competitive advantage over others; and instead of signing a 5-year contract after the competitive bidding the Assembly engaged the companies for 3-months probation through letter of acceptance with the intention of replacing some of the companies with those who have political connection.

The concerns raised by the key informants are real issues which show that regulatory rules are not fully applied in solid waste management due to the weaknesses in the institutional and regulatory arrangements and therefore low cost recovery and poor service quality problems are bound to exist.

8.4.2   Regulatory Policy and Control Instruments

Regulatory theory stipulates that businesses are regulated to balance the interest of the regulated businesses with the public interest (Balance and Taylor, 2005; Franceys and Gerlash, 2008). Regulation is achieved through regulatory instruments such as legal and financial incentives for the service delivery. The legal and financial incentives in the Ghana case are the promulgation and enforcement of by-laws, regular fixing of service charges and collection fees and the award of signed contract which are the constitutional mandate of the Assemblies. The interest of the private companies is to make profit and therefore they look for financial incentives in any regulatory arrangement. The provision of financial incentive is seen in terms of regular payment of contract, full cost recovery service charges and collection fees, provision for adjustment of contract price and service charges to reflect inflation, payment of interest on delay payments and long duration of contract. In this study, the findings show that there was lack of financial incentives and this in turn has resulted in weak capacity of most of the companies. It takes a long time (over 2 years) to adjust service fees to reflect inflation. Again, it takes a long time (over a year) before part of delayed payment is made. Even though earlier contracts state that in the event that payment is delayed beyond three months, the company shall be paid interest, most of the companies said no interest is paid on payment delayed. These results are consistent with the findings of Awortwi (2003:185).

Literature on contract theory stipulates that robust competition, tight contract specifications, rigorous contract monitoring, and the use of legal means to enforce the contract are factors that are supposed to work together in contractual relationship and to exert control over the contractor. The findings of Fernandez (2005) suggest, however, that successful contractual relationships are not discrete, arms-length transactions that are programmed and controlled but rather dynamic relationships that are managed in a flexible and cooperative manner (relational contracting), with the parties acting more like equals than principal and agent. The significance of trust and cooperation as determinants of contracting performance raises some concerns about loss of accountability in contracting for services (Fernandez, 2005), since the external mechanisms of accountability, such as oversight and regulation, are needed to hold private contractors accountable for their behavior and service delivery. In the case of Ghana, the regulation by contract is dynamic and relational such that trust and cooperation of the private sector play role in solid waste service delivery, but there is power relation imbalance as the principal (Assemblies) has more control over the agents (companies).

8.4.3 Regulatory Strategies and Incentives

There are some necessary strategies for improving solid waste collection and management. First, there is the need to separate the role of regulation from service delivery. The municipal authority could form a company to replace the in-house delivery to allow the local government to regulate better and obtain cost data and service delivery standards for monitoring. In this case the hidden cost and real cost of running a company could be obtained from the public company.

Second, the demarcation of zones for competitive bidding as of now is not done to include low, middle and high income areas to afford cross-subsidisation. The zones which were mainly low incomes areas were not bided for in the Accra competitive bidding. The current demarcation along administration sub-metro boundaries is not the best for competitive bidding for all zones. Each sub-metro could be divided and mixed sub-units of sub-metros should form a zone. This will create the necessary competition for the market.

Third, the capacity for enforcing by-laws and sanctioning defaulters is weak. The Environmental Health Department (EHD) and the Waste Management Department have different directors and therefore the coordinating effort to enforce by-laws is weak. Cases reported by the solid waste and sanitation unit to the environmental health unit are not followed-up and defaulters are not prosecuted. The way forward is to restructure the existing institutional arrangement and place an environmental health unit for sanitation and solid waste under the director of Waste Management Department. The environmental health unit for food hygiene and disease control could be placed under the existing director of Environmental Health Department. This will provide the needed support and effort to enforce the by-laws. Another option is to maintain the existing arrangement but second some staff of the Environmental Health Department to work with the companies since EHD has the legal right to do so. With this arrangement they will have the necessary logistics and support to prosecute offenders and defaulters of service charges, and report to their department.

The enforcement of service standards in contracts is constrained by local governments (Assemblies) weak financial capacity and low cost recovery. The way forward is to put in more effort to implement the policy on full cost recovery in the environmental sanitation policy of 1999 and 2007, and to make the companies more responsive to users or service recipients. The companies should be tasked to collect user charges and to involve the service recipients in the monitoring of service quality. The people, when given a 'choice' and 'voice' in the service delivery and quality monitoring, will provide the incentive to pay for the service and will compel the companies to provide better services. In development theory, user involvement as a process or means by giving the people 'choice and voice' ensures that the desired outcome of service quality is met (Awortwi, 2003:88, 89).

8.5   Conclusions

This chapter examined the contractual arrangements and regulation of private sector involvement in solid waste collection using Ghana as a case study. This chapter focused on the regulation of the private sector involvement in solid waste collection in five cities in Ghana. The regulation was assessed using: the contracts' scope and specifications; regulation of user charges, quality of service, and access to information; monitoring of companies compliance with regulations and supervision by the Assemblies; and Assemblies adherence to the regulations in the contracts signed with the companies. The data collected were analysed and the bottlenecks of the contract, legislation, and financial arrangements for involving private companies in solid waste collection were explored. Five finding were identified. First, there was a shift from unsigned contract arrangements (giving of letters to notify company to render service or to continue service without signed contract) to providing opportunity for companies to renew contracts through competitive bidding. The research revealed that there were regulatory arrangements (contractual and financial) in place for the involvement of the private sector in solid waste collection, but there are weaknesses in the arrangements. One conclusion drawn from the research was that previously, the companies did not undergo competitive bidding when their contracts expired but now there is a shift toward competitive.

Second, the obligations of the Assemblies in the contracts are in practice not strictly adhered to by the Assemblies. The adherence to contract obligations, enforcement of by-laws and sanctioning of companies for non-compliance with terms and conditions in the contracts were challenges. The by-laws on solid waste and sanitation are available to sanction offenders but they are not strictly enforced. The sanctions on companies were usually not enforced because the Assemblies most of the time failed to adhere to their obligations of paying the companies within 45 days of claim submission.

Third, there was an element of regulatory capture in the contracts signed between one of the companies and the Assemblies. Most of the companies admitted that there were delays in payments by the Assemblies. The payment of the contracts delays for more than a year due to inadequate internally generated funds from user charging (low cost recovery) and delay in the disbursement of District Assemblies Common Fund from the central government. Whereas subsidies payments delayed over a year, one company received regular payments within a year. This company that received regular payments had political connections with the politicians in government at that time. The regular payment to this company could be attributed to 'regulatory capture' by interest groups in government with rent-seeking and collusion behaviour. There was a repeal of the clause on payment of interest on delay in payment. This is associated with 'agency capture' regulation which seeks to avoid interest payment on monies delayed over a year.

Fourth, efforts are being made to collect user charges for the communal mode of service using pay-as-you-throw mechanism. The involvement of the private sector in solid waste collection has increased the collection rate and the proportion of house-to-house collection service without subsidy from the government, compared to when the Assemblies were rendering the service alone. The user charges vary from city to city and also vary within the same city because of the income levels of the people, but some of the companies served the areas where the user charges were low, and this may lead to low profit and inadequate private finance. The companies perceived that the user fees set are not reasonable to recover cost especially the medium and low class residential areas.

Fifth, the obligations of the companies in the contracts are in practice not strictly adhered to. The companies have reduced house-to-house collection frequency from two to one, and the repairs of communal containers are not done. Trust, cooperation, and decrease in the quality of service appeared to be the driving factors for sustaining the private sector in business, but the companies were of the view that they were highly responsive to the Assemblies' requirements of service quality.

The analysis shows that the regulation of service charge and unit price by the Assemblies and regulatory capacity influence the productivity of the companies. The ability of the Assemblies to set appropriate charges and make regular payment of waste collected was perceived to be inadequate by the companies. It can be concluded that the differences in performance (productivity) of the companies were purely due to the differences in the internal capacities of the various companies and the inadequate regulation and capacity of the Assemblies. The Assemblies did not comply with their obligations since all the Assemblies had weaknesses in the regulation of solid waste services and defaulted in subsidy payment and payment of interest on delayed claims. The major challenges of solid waste collection were the inadequate physical capacity of the Assemblies, the long delays in paying for the contracts, limited funds from central government, low cost recovery, inadequate monitoring of quality of service, and inadequate human capacity for planning and implementing solid waste management programmes. There was evidence of limited capacity for regulation from the key informant interviews.

9 Conclusions and Implications

9.1 Introduction

This thesis focused on the private sector involvement in solid waste collection, and in particular the private sector performance and capacity and local governments' regulation of the private sector. There are gaps in the literature on the understanding of the private sector involvement in solid waste management in developing countries, and the factors which influence private sector performance and participation which this research has tried to address. It is useful to understand the factors why certain private sector companies are more efficient than other companies since certain companies have competitive advantage over others. When the factors for higher private sector efficiency are understood, it is possible to prepare arrangements for involving private companies such that higher efficiency is achieved in practice (Cointreau-Levine and Coad, 2000).

It is believed that private sector involvement in urban solid waste collection services seeks to improve service efficiency, service quality and increase investment. However, if the inter-organisational arrangements and regulatory practices are not well structured to provide incentives, private sector involvement may not achieve these objectives, regardless of the market forces (Cointreau-Levine and Coad, 2000). Besides, private sector companies cannot always be more efficient without a real effort to apply management techniques (Cointreau-Levine and Coad, 2000). The literature on service quality and efficiency gains by the private sector delivery over

284

that of the public sector delivery shows that the results of internal and external factors influencing private sector performance are inconclusive (Carter et al., 1992; Post et al., 2003; Bel and Coasta, 2006; Bel and Warner, 2008). This study explored the explanatory factors of private sector efficiency and service quality.

The focus of this study was on the performance differences among private sector companies. Within the private sector at the firm level, theory predicts differences in efficiency and service quality. In this thesis, the focus was on the firm level efficiency and service quality, and the factors that explain the private sector performance in urban solid waste collection. The main objective of this research was to examine the evolving involvement of the private sector and households in urban solid waste collection, and the factors explaining differences in performance of private sector companies. The main research question is: To what extent do capacity and regulation for solid waste provision influence private sector performance? The sub-research questions of this study are: How has private sector involvement evolved? What is the level of company performance (productivity and service quality)? What factors explain the level of performance? And what are the drivers for improving performance of service delivery? The issues of efficiency and service quality and the explanatory factors were explored by focusing on private sector capacity (scale of operation combined with certain skills/capabilities) and regulation practices (contracts, competition, accountability for results, political interference) and their implications for urban solid waste collection.

The results of this study have been presented in Chapters 5 to 8. The study involved the analysis of data obtained from public sector organisations (Waste Management Departments in five cities), a survey of a total of 25 private sector companies and a survey of 1500 households across the five cities. *The main hypothesis of this study states that private sector companies, with adequate capacity for service provision, are likely to achieve high performance when strong regulation practice exist and when formal rules and contractual obligations are strictly adhered to.* The specific sub-hypotheses, based on the overview theory, that were used to guide this study are that:

o   High productivity of a company leads to better service quality, and
      that high performance is achieved when both productivity and ser-
      vice quality are high.

o   The increasing involvement of the private sector coupled with
      households' involvement in user charging, service quality monitor-
      ing and waste minimization programme leads to customer-oriented
      service and better service quality.

o   Strong regulatory practices and adherence to formal rules and con-
      tractual obligations provide incentives for cost recovery and better
      service quality.

o   Scale of operation of private companies and the capability for solid
      waste collection influence private sector performance (productivity
      and service quality).

This chapter presents the main findings on the performance and the factors
that influence private sector performance. It also provides conclusions and
implications for theory and practice. The chapter is divided into 5 Sections.
Section 2 presents the research findings on private sector performance,
while Section 3 presents findings on the explanatory factors (evolving involvement processes, capacity and regulation) influencing private sector
performance. Section 4 presents the research conclusions drawn from the
findings and the hypotheses. Section 5 draws the theoretical implications
of the findings of this study as well as areas for further research, and implications for policy and practice.

9.2   Private Sector Performance in Solid Waste Collection

The first research question is: What is the level of company performance (productivity and service quality)? The results on private sector performance in chapter 6 show that there were variations in productivity, technical efficiency and service quality of companies within and across the five cities. Results show that 17 out of the 25 companies had technical efficiency values below 50%. Whereas 7 companies in Accra, Kumasi and Tema performed above average, all the companies in Takoradi and Tamale performed below average. The reasons for the low productivity and inefficiencies are attributed to inefficient operations, low utilization of vehicles and sub-optimal scale of operation of the companies. Companies with vehicle productivity above average had higher average daily quantity of waste collected (in tonnes per vehicle) for both collection methods than those companies with productivity below average.

Furthermore, the analysis suggests that there was no significant correlation between service quality and productivity. There was also no significant difference in the service quality of below and above average productivity performers. These mixed results of service quality were due to either the service demands (customers) served by the companies were less (diseconomies of scale) or quantities of waste generated were more than the physical capacities of the companies. This finding is contrary to the hypothesis that high productivity of a company leads to better service quality, where high performance is achieved when both productivity and service quality are high. This study shows that high productivity and technical efficiency do not necessarily lead to higher service quality, due to the weak capacity and regulation of the private sector. It suggests that service quality is compromised if the capacities of the companies cannot cope with the increasing waste generation even though high productivity and technical efficiency may be achieved. The high productivity was due to profit motive of companies. As long as the users are not empowered, the companies will reduce service quality to make profit.

There was significant difference in service quality across cities, and the service quality of house-to-house collection was higher than that of communal collection. The residents' perceived service quality across the five cities showed that service quality for house-to-house collection was slightly higher than communal collection. In communities where there was regular collection of waste generated and no waste overflow, the residents' rating of service quality was high, whereas in communities where there was irregular collection and waste overflow the rating was low. Service quality for house-to-house was high because residents pay for the collection service and as a result the collection by the companies was regular as scheduled and reliable. This result is consistent with findings from Awortwi (2003:209) that house-to-house service quality was higher than that of communal collection. Solid waste collection is a visible service and residents are better able to rate the perceived service quality (Batley, 2001). The level of service (house-to-house and communal), subsidies and payment of user charges could influence the perceived service quality observed across the cities.

9.3   Factors Influencing Private Sector Performance

9.3.1   Evolving Practice of Private Sector Involvement

Private Sector Involvement

The second research question is how has the private sector involvement evolved over that past two decades? The analysis of the evolving practices of the private sector and households' involvement presented in chapter five show that there is an increasing role of the private sector in solid waste collection service over the past two decade with each company operating in a defined collection zone. More than 60% of the solid waste in the cities was collected by the formal private sector. The involvement of the private sector started with five companies in Accra and six in Tema rendering franchised house-to-house service in the early 1990s. The involvement of private sector was an exogenous initiative of external support agencies. There were no competitive bidding and contracts but the companies were given letters to commence solid waste collection without subsidy. The first competitive bidding for franchise house-to-house solid waste collection started in 1997 in Tema as part of the World Bank's project on Urban Environmental Sanitation. The second competitive bidding happened recently in 2008 in Kumasi and in 2009 in Accra. All communal services were rendered under service contract with subsidy, whereas franchise house-to-house service was rendered with or without subsidy and with or without contract. However, in recent development there has been a shift from separate service contract (for communal) and franchise contract (house-to-house service) to collection contract covering both house-to-house and communal service in a single contract.

Implementation of Policy on Private Provision of Services

According to the Ghana Environmental Sanitation Policy (1999), the Waste Management Department of the Assembly is mandated to maintain an in-house capacity to provide twenty per cent (20%) of the solid waste services directly and outsourced the 80% to the private sector. The public sector in Tema, Tamale and Takoradi collected 22%, 26% and 45% of solid waste respectively, whereas in Accra and Kumasi there was no direct public provision. The degree of private sector involvement depends on the size of the city and the in-house physical capacity of the Assembly to provide direct service. The Assembly with limited in-house capacity requires private sector with adequate capacity to increase collection rate.

There was slow pace in the implementation of the cost recovery policy prior to 2008, but it has been taken on board recently together with competitive bidding required by the Procurement Act. The involvement of the Assemblymen (and Assemblywomen) in the collection of illegal user charges and lack of political will derailed the full scale implementation of cost recovery until recently. Even though there was pay-as-you-throw (PAYT) in Accra and Takoradi, there was no accountability to the Assemblies by those collecting the money. The low cost recovery has led to financial constraints (Assemblies still owe companies) and therefore Metropolitan Assemblies are unable to keep and maintain the cities clean. The Assemblies and the companies are however faced with the problems of inadequate waste collection equipment and containers/bins, lack of funds to procure new truck, lack of political will to support cost recovery of collection service, and unwillingness of some residents to contribute towards cost recovery of waste collection.

Cost Recovery of Service Provision

A shift towards cost recovery to charge end users on pay-as-you-throw (PAYT) basis for communal service and monthly flat charges for kerbside collection began recently with the city-wide-fee for solid waste collection in Accra, Kumasi and Tema. Effort to involve more households to pay for service began in 2008 in Kumasi and 2010 in Accra. About 50% of the 150 container sites in Kumasi were under PAYT at the start of the communal service cost recovery. The pay-as-you-throw (PAYT) charging mechanism introduced in Kumasi suggests that there is potential for full cost recovery since the analysis show that 60% of the solid waste collection cost is paid by PAYT revenue. With proper management and the cost accounting, PAYT will provide a source of revenue to replace subsidies being paid to the private companies. There is potential for reduction of subsidy and full cost recovery for solid waste collection if households are involved in the service provision. The PAYT mechanism, even though it failed in Accra in 1995, is working well in Kumasi. The success of PAYT depends on public awareness, involvement of the households, and strict enforcement of the by-laws on illegal/indiscriminate dumping.

The approach to achieve full cost recovery in Accra through city-wide-fee-based solid waste collection was different from that of Kumasi. The PAYT approach was used in Kumasi for the low income communities, whereas in Accra a monthly service fee was imposed on low income households without giving the people 'choice and voice'. The implication is that the poor people are served with higher service level (house-to-house collection) which is costly compared to communal collection. PAYT affords the households to reduce cost through waste reduction and will cost them half of what they would have paid for house-to-house collection. This study shows that PAYT where could be a better option of user charging for low income communities, instead of a monthly flat fee per household. PAYT does not accumulate the charges compared to the monthly fee. The problems with the fee-based collection in Accra are: (1) financial burden on the companies to distribute bins to all households, (2) unwillingness of households to register for collection, (3) low payment rate at the low income communities, (4) illegal and indiscriminate dumping by some households who refused to register, (5) high disposal fee (tipping fee) to be recovered from user charges, (6) weak enforcement of illegal dumping, and (7) weak enforcement of registration by the Assembly. The introduction of city-wide fee-based collection of user fees at household levels in Accra is intended to raise revenue to cover both collection and disposal cost without subsidy from Assembly, but there could be a gradual withdrawal of subsidy for the poor communities as it is the case in Kumasi.

Households' Perception on Cost Recovery and Willingness-to-pay

Results show that 70% of the respondents across cities who were served with house-to-house collection were of the view that solid waste generators and Assemblies should pay for the cost incurred in waste collection and disposal. For the communal collection, 55% of the respondents across cities were of the view that cost should be paid by generators and Assemblies whereas 39% said the Assemblies should pay for the collection cost. This presuppose that more effort has to be put into household involvement in the collection service and the companies made to be accountable for results and be more responsive to the user. The involvement of the households to pay for the service and the increase in the level of awareness helps to make the private sector more responsive to the customers to realise the policy on full cost recovery.

The willingness-to-pay more to improve house-to-house service was low as a result of their perceived dissatisfaction with the current service quality.

Some respondents (20%) were willing to pay current charges for the service, while 62% of respondents were willing to pay less than what they are paying. The respondents were demanding value for money and want to see improvement in service delivery first before they will pay more. They said the user charges should be based on collection frequency (twice collection), waste volume, and regular collection instead of the monthly fixed fees. The implication is that increasing private sector involvement without corresponding increase in households' involvement does not ensure that private sector becomes responsive to the customers and to provide better service quality. This means that customer orientation and accountability for results and better service quality are the factors for increasing service charges. The private sector needs to be responsive to the customers and to provide better service quality, and this need is affirmed by some authors (World Bank, 1997; Gidman at al., 1999).

The results of willingness-to-pay for communal collection service show that about half of the respondents were willing to pay for the service. The results show that 53% of the respondents across the cities at the time of the survey did not pay. The survey shows that 47% of the respondents were paying-as-they-throw waste into communal containers. These were illegal collection of user charges in Accra (92%) and authorized but unaccounted-for revenue collection in Takoradi (70%). The collection of the charges at container points in Accra was illegitimate, since the collection of user charges was initiated by some Assemblymen but not approved by the Assembly.

Households' Perception on Source Separation, Waste Recycling and Service Quality Monitoring

The results show that 48% of the respondents across the cities were willing to source separate waste. The respondents in Kumasi, Tamale and Accra who were willing to separate organic waste for separate collection were 91%, 68% and 55% respectively. Some of the respondents knew the benefits of composting. Those who were willing to separate indicated a number of incentives needed for separating waste for collection. These include provision of free polythene bags of different colours, increase collection frequency to twice a week, some agree to buy their own bins for storing organic waste and pay reduced fee, and some want free bins to be provided by companies and assembly for separate collection. Some suggested that there must be educational campaign and that the separation be made by-law to compel everybody to separate. The education and motivation of households could play a role in creating the necessary awareness about source separation of waste.

The respondents were aware of informal waste reuse and recycling. Most of the respondents said they reuse empty plastics and cans and sell plastics containers and sachet bags/plastics to buyers and scavengers, and burn combustibles. Despite the reuse and recycle of these waste fractions, greater amount of the reuse and recyclable materials are found in the mixed waste stream which eventually are contaminated and find their way to the final disposal site. There is no formal arrangement for source separation and collection of these recyclables. The companies have no plans to undertake source separation for waste recycling since there are no economic and legal incentives to do so at the moment. With the initiative of city-wide-fee for solid waste collection, there will be the need to enact legislative instrument to regulate recycling and reduce organic fraction of waste sent to landfill as well as allowing private sector to own and manage landfill privately. This will provide the necessary incentives for private sector involvement in waste recycling and resource recovery to ensure sustainable development.

Most of the respondents were aware of people not served and the indiscriminate dumping of waste and measures to prevent the illegal dumping. The preventive measures suggested by the households include collection service improvement, enforcement of by-laws, and public education. The study revealed that there was no complete coverage of all households in communities where house-to-house is served, since some of the households (16%) opted out and were not served in areas where companies operate, especially in the middle income areas.

The households' involvement in service quality monitoring has not received the needed attention. More than half of the respondents of the house-to-house service were not happy about the quality of service. They mentioned that waste was uncollected for a week or more, and the collection frequency was not consistent resulting in waste overflow and littering. There are no formal channels and arrangements for service quality monitoring, customer complaints and responses to complaints, as required by customer orientation. The respondents attributed the problems of overflowing of waste from skip containers onto the ground at the communal sanitary sites to irregular collection of containers, no watchmen at night or lack of attendants at container sites, poor positioning of skip containers by truck drivers, inadequate support for children who send waste to containers, and negative and lax attitude of inhabitants. The environmental problems and the lax attitude and behaviour of the people can be dealt with through public education and accountability of the companies for results.

Weak Involvement of Households

It was hypothesized in this study that increasing involvement of private sector coupled with households' involvement in service provision leads to customer-oriented service and better service quality. The findings from this study partially support this hypothesis, due to the slow involvement of households over the past two decade. The involvement of households in paying for services, service quality monitoring, source separation, waste recycle, and attitudinal and behavioural change has been slow. The new public management theory suggests that customer orientation allow for collection of fees towards cost recovery (van Dijk, 2004). The involvement of more households to pay for services is a recent development in Accra, Tema and Kumasi. The household involvement should be on all elements of sustainable solid waste management covering not only paying for services but also service quality monitoring, source separation, waste recycle, and attitudinal and behavioural change.

9.3.2   Regulatory Factors Influencing Private Sector Performance

The third research question is: Do regulatory practices explain private sector performance? The analysis of the influence of regulatory practices on private sector performance is presented in chapter 8. Regulation of private sector involvement was assessed based on formal rules and actual practices of contracting-out (Assemblies adherence to regulations and signing of contracts with the companies) and the extent of regulatory control in practice (regulation of user charges, quality of service, monitoring of companies compliance to contract), and supervision by the Assemblies. The bottlenecks of the existing regulation, contract and financial arrangements for involving private companies in solid waste collection were explored.

Five findings on the analysis of regulatory practices and adherence to formal rules and obligations are drawn from this study. First, there was a shift from unsigned contract arrangements (giving of letters to notify company to render service or continue service without signed contract) to arrangements involving competitive bidding and cost recovery from user charging. The research revealed that there were regulatory arrangements (contractual and financial) in place for the involvement of private sector in solid waste collection. Each company operated in a particular delineated zone (no open competition allowed) and some contractual agreements (with or without subsidy) existed between companies and Assemblies. One conclusion drawn from the research is that the companies did not undergo competitive bidding when contract expired but now there is a shift toward competitive bidding. There is a gradual well functioning system being put in place with the recent competitive bidding, signing of contracts, and city-wide user charging.

Second, some regulatory rules and the obligations of the Assemblies in the contracts were, in practice, not strictly adhered to by the Assemblies. The adherence to contract obligations in terms of enforcement of by-laws, and payment of subsidies was not adhered to. There was an element of 'regulatory capture' in the service provision since some companies had contracts signed whereas others did not. The contracts were signed by the Metropolitan Chief Executives, who are government appointees, without going through the Assemblies Tender Boards. Some companies had links with the government in power and had contract with the Assemblies while the other companies in Accra and Tamale had no contracts. The Assemblies (municipal authorities) regulate service charges with or without subsidy for franchise house-to-house service. The unit price per tonne used to pay the companies for communal collection is set by the Assemblies alone. The upward review of unit price and service charges was always delayed by the Assemblies for more than two years. Besides, the payment of subsidy for the communal collection delays for more than a year but no interest is paid on money delayed. However, one company received regular payments and this is an indication of regulatory capture. By-laws on solid waste and sanitation are available to sanction offenders but are in practice not strictly enforced. This company did not perform better than the rest. The regular payment to this company could be attributed to 'regulatory capture' by interest groups in government with rent-seeking and collusion behaviour. The company had political connections with politicians in government at the time. There was a repeal of the clause on payment of interest on delay in payment. This can be associated with 'agency capture' regulation which seeks to avoid interest payment on monies delayed over a year. However, efforts are being made to collect user charges for the communal mode of service using pay-as-you-throw mechanism and fixed service charges.

Third, the obligations of the companies in the contracts are, in practice, not strictly adhered to. The sanctioning of companies for non-compliance with terms and conditions in the contracts was a challenge. The sanctions of companies were usually not enforced because the Assemblies most of the time failed to adhere to their obligations of paying the companies within 45 days of claim submission. The companies reduced house-to-house collection frequency from two to one, and the repairs of communal containers were not done. Trust, cooperation and decrease in the quality of service appeared to be the driving factors for sustaining the private sector in business, but the companies were of the view that they were highly responsive to the Assemblies' requirements of service quality. There was evidence of waste overflows but sanctions for non-compliance were not enforced because of delays in payment.

Fourth, the analysis shows that regulation of service charge and unit price and managerial autonomy of the Assemblies influence productivity of the companies. The ability of the Assemblies to set appropriate charges and make regular payment of waste collected was perceived to be inadequate by the companies. It can be concluded that the differences in performance (productivity) of the companies were due to the differences in the internal capacities of the various companies, inadequate regulation and inadequate capacity of the Assemblies. It can be concluded that Assemblies' obligation, financial autonomy and regulatory capacity did influence productivity of companies, since all the Assemblies had weaknesses in the regulation of solid waste services and defaulted in subsidy payment and payment of interest on delayed claims.

Finally, the major challenges of solid waste collection were the inadequate capacity and resources of the Assemblies and regulatory capture by politicians. There were long delays in paying for the contracts, poor monitoring of quality of service, limited funds from central government, low cost recovery, and inadequate human capacity for planning and implementing solid waste management programmes. The Assemblies are not well staffed with the right personnel for the work. The ability and competence to collect data, analyse and take evidence-based decisions require high level training, possibly master's degree, but this is lacking in four cities.

9.3.3   The Importance of Private Sector Capacity for Good Performance

The fourth research question is: Does the capacity of the private sector explain the level of performance? The results of private sector capacity in chapter 7 suggest that the size of a company significantly explained partly the variation in productivity. The productivity of private companies' increases with increasing scales of operation (size of company) and decreases beyond optimum scale of operation. The study suggests that companies with 5 to 7 vehicles seem to have optimum scale of operation, for the purpose of economics of scale. It can be said that scale of operation of a company matters in solid waste collection.

The productivity of companies was influenced by supervision capability. There were mixed results of capability for collection planning, in-house maintenance capacity, operation scheduling, and workers morale. It can be concluded that optimum scale of operation of a company and supervisory capability of a company lead to high productivity. The findings confirm the hypothesis that scale of operation and supervisory capability of private company influence productivity. The inefficient scale of operation and inefficient supervision explain the inefficiencies of the poor performing companies across the cities. The analysis shows that scale of operation (size of company), companies' experience and supervisory capability for solid waste collection lead to increasing high productivity.

9.4   Research Conclusions

The study on private sector involvement in solid waste collection, and in particular private sector's capacity and local governments' regulations of the private sector bring to the fore the factors which influence private companies' performance in terms of productivity and service quality. Six conclusions are drawn from the findings of this study. First, the conclusion drawn from the findings on performance is that there was a significant difference in service quality across cities, and that the service quality of house-to-house collection was higher than that of communal collection, but there was no significant correlation between service quality and productivity. This study suggests that high productivity of the private sector does not necessarily lead to better service quality due to profit motive, capacity, and regulatory factors influencing performance. The finding of this study is contrary to the hypothesis that high productivity of a company leads to better service quality, and that high performance is achieved when both productivity and service quality are high. This study suggests that service quality is compromised if the capacities of the companies cannot cope with the increasing waste generation even though high productivity and technical efficiency may be achieved as in the case of this study.

Second, there has been an increasing role of the private sector in solid waste collection service over the past two decade with each company operating in a defined collection zone. More than 60% of the solid waste in the cities was collected by the formal private sector. The practices of involving the private sector have been the delineation of city into zones, engagement through competitive bidding or no bidding, and rendering of communal or house-to-house collection service with or without signed contractual agreements between companies and Assemblies. There is a gradual well functioning system being put in place with the recent competitive bidding, signing of contracts, and city-wide user charging.

Third, notwithstanding the increasing involvement of the private sector, the involvement of households in paying for service provision, service quality monitoring, waste recycle and source separation has been slow. The Assemblies and companies have not ensured customer-oriented service due to the slow pace of households' involvement. Cost recovery is low in communities where there is low level of households' participation in the service provision. Most of the people are aware of the preventive measures of illegal dumping, and these are collection service improvement, enforcement of by-laws, and public education, to deal with the solid waste problems.

Fourth, a shift towards cost recovery through charging all households on pay-as-you-throw (PAYT) basis for communal service and monthly fixed charges for kerbside collection (house-to-house) started first in Tema and recently in Kumasi and Accra. The approach to achieve full cost recovery in Accra was different from that of Kumasi. The PAYT approach with potential for full cost recovery was used in Kumasi for the low income communities whereas in Accra a monthly service fee was imposed on low income households. The companies involved in the fee-based collection in Accra are faced with problems of financing bins, unwillingness of households to register for collection, low payment rate at low income communities, illegal dumping by some households who refused to register, high disposal fee, and weak enforcement of registration by the Assembly.

Fifth, there were weak regulatory practices and non-adherence to contractual obligations and these consequently provided no incentives for full cost recovery and better service quality, but now a gradual well functioning system is being put in place. The regulatory practices and 'capture' which affected service quality are unsigned contracts, prolonged periods before upward review of collection fees and service charges, no interest on monies delayed, and delay in payment of subsidy. The delay in subsidy payment (for more than a year) to the private companies does not provide incentive mechanism for private sector growth and does not enhance better service delivery. Results suggest that 'regulatory capture' exists in contracting since some companies had contracts signed and received regular payments whereas others did not. There was weak adherence to obligations such as sanctioning of private companies for non-compliance and enforcement of by-laws. The weak enforcement of legislation does not provide sufficient incentive mechanisms for better service quality and full coverage. Regulation of prices by the local governments, where the levels of service charges and collection fees set are not sufficient to cover full cost of service, undermine the service quality of the companies. The major constraints of solid waste management were the inadequate capacity and resources of the Assemblies and the regulatory capture by politicians.

Sixth, the results suggest that companies' experience, scale of operation, and supervision capacity lead to high productivity. The productivity of private companies increases with increasing scale of operation and decreases beyond optimum scale of operation of 5 to 7 vehicles. The study confirms the hypothesis that the scale of operation of private companies and the capability for solid waste collection influence private sector performance (productivity and service quality). The results of the influence of factors such as route planning, operation scheduling, workers morale, and in-house maintenance on productivity of the companies were mixed and inconclusive. The inefficient scale of operation and supervision of operations explain the poor performance of companies.

9.5 Implications for Theory, Practice and Further Research

9.5.1 Theoretical implications

This section presents the theoretical contributions of this study to the research and theories on private sector involvement in public service delivery. This research is aimed at contributing to the understanding of regulation, private sector capacity and performance within the context of new public management (contracting-out public services, user charging, and the role of customer orientation in service delivery). The analysis of the private sector involvement in solid waste collection has provided findings and understanding of the evolving private sector involvement, private sector capacity and local governments' regulations of private sector and the factors that explain private sector performance in the Ghanaian context.

The implications of the findings from this study are discussed with respect to the four hypotheses and theoretical claims in the literature review. To examine the theoretical implications of this study, the theoretical themes presented in chapter 3 are used as lens to view and examine the findings of this study and to raise theoretical issues for further research and discussions. The four theoretical themes to help understand the private sector performance in the local context and within the scheme of study are: performance, private sector capacity, regulation, and user involvement in private provision of public services. The theoretical implications drawn from the findings of this study are discussed these themes.

Private sector performance

It was identified in this study that there were variations in productivity, technical efficiency and service quality of companies within and across the five cities. The comparison of the actual productivity with the maximum potential productivity shows that most of the companies were performing poorly. Productivity is a relative concept and its measurement requires a standard of performance against which the success of economic units is assessed (Mayes et al., 1994). Productivity indicates the success with which companies are best utilising their available input resources to produce the maximum potential output. The maximum potential productivity could be either the best actual performance in an industry or a theoretical maximum (Mayes et al., 1994). In theory, technical efficiency is the extent to which companies either use excessive input to produce their output with a given input mix, or under-produce output given their input and mix. For instance, the companies in Tamale under-produced given their input mix.

Performance is composed of productivity, profitability (cost recovery) and service quality. Productivity is the central core of performance and has operational definition as the relation between output quantity (solid waste collected) and input quantity (collection vehicle). The combination of high values of efficiency and effectiveness (productivity and service quality as used in this study) in the delivery process leads to high performance. The first hypothesis for the study states that high productivity of a company leads to better service quality, and that high performance is achieved when both productivity and service quality are high. In theory, a well performing company always strives to achieve high productivity and provides better service quality. It is possible for an effective organisation to be inefficient and an efficient organisation to be ineffective (Tangen, 2005). This study suggests that high productivity of the private sector does not necessarily lead to better service quality. An implication of this finding is that for a company to be rated as a well performing one, it has to achieve high productivity and service quality as the theory suggests. The companies cut service quality to make profit and get away with poor performance.

Evolving practice of private sector and user involvement in service provision

The second hypothesis for the study states that increasing involvement of private sector coupled with households' involvement in service provision does ensure customer-oriented service and better service quality. The finding in this study was that the increasing involvement of the private sector was not commensurate with increasing households' involvement in service provision resulting in weak customer-oriented service and poor service quality. There is the need for improvement in customer orientation for the house-to-house recipients. The willingness-to-pay more was low as users demanded value for money. In theory, for those who are paying for service, inadequate service quality perceived by users is disincentive to people to pay more for services. The theoretical implication is that for the companies to increase service charges, they first of all have to orient the customers and be responsive to improve service delivery.

Despite an increase in private sector involvement in communal service contracts with subsidy, households were not involved in paying for service and service quality monitoring. The service quality of the house-to-house was better than that of the communal service and house-to-house service contract with subsidy. In theory, households' involvement in service provision in terms of choice of service level, paying for service, and reporting complaints of poor service ensure customer-oriented service and better service quality (Gidman et al., 1999). Besides, households are given 'choice and voice' by involving them in choosing of preference for service level, paying for services, and monitoring of service quality. The extent of household involvement has ramification for cost recovery and service quality. In theory, cost recovery is low in communities where there is low level of households' participation in the service provision, and this was the case for the communal collection service.

There is a shift after the enactment of procurement act, which requires competitive tending and signing of contract, from non-involvement of households to now involving households to pay for the service to reduce the financial burden on the local Assemblies. Effort to involve more households to pay for service began in 2008 in Kumasi and 2010 in Accra. The pay-as-you-throw (PAYT) charging mechanism introduced in Kumasi suggests that there is potential for full cost recovery.

Figure 9.1 provides a typology of customer involvement (high or low) and service quality (high or low) for solid waste collection. The current regulatory practice is I (for communal service) where cost recovery is low and payment is delayed, but there is a shift to IV where all service recipients are required to pay to improve service quality (Figure 9.1). The current practice for house-to-house service is III and is shifting to IV since the Assemblies have now combined both services into one contract with more focus on cost recovery through user charging, contract monitoring, and enforcement of sanctions.

Involvement of Households in paying for service

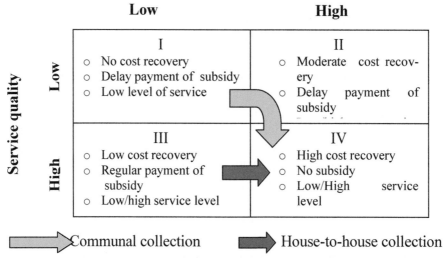

		Low	High
Service quality	**Low**	I ○ No cost recovery ○ Delay payment of subsidy ○ Low level of service	II ○ Moderate cost recovery ○ Delay payment of subsidy
	High	III ○ Low cost recovery ○ Regular payment of subsidy ○ Low/high service level	IV ○ High cost recovery ○ No subsidy ○ Low/High service level

⇨ Communal collection   ➡ House-to-house collection

Figure 9.1 Evolving typology of customer orientation-service quality relationship

The pace of development of solid waste services depends on the level of public awareness of solid waste-related issues and user participation in making improvements happen at the ground level (Wilson et al., 2001). The public support for any issue can be greatly increased if the public is well-informed of the reasons behind the actions and the intended benefits. The areas for public awareness and participation include public cooperation for service level provided, paying for collection service, involvement in service quality monitoring, and involvement in recycle and source-separation activities. The involvement of users in the private provision of public services has implications for the service quality and the level of cost recovery. Also the literature on participation and customer orientation viewpoints shows that the involvement of users in service provision to pay for service and report complaints on service quality has ramifications for full cost recovery and better service quality.

Cost recovery is stated in the policy that solid waste shall be subsidised until the time that full cost recovery could be achieved. The policy on cost recovery did not address the core mechanisms and strategies. However, efforts are being made to collect user charges for the communal mode of service using pay-as-you-throw mechanism.

Regulation practices and adherence to formal rules and contractual obligations

The third hypothesis for the study states that weak regulatory practices and non-adherence to formal rules and contractual obligations are disincentives for cost recovery and better service quality. The research revealed that there were regulatory arrangements (contractual and financial) in place for the involvement of private sector in Solid Waste Collection. The national Procurement Act makes it mandatory for the Assemblies Tender Boards to use competitive bidding to select companies. The environmental sanitation policy, Local Governments By-law Act, Local Government Act and other national legislative instruments are in support of solid waste management. These formal rules (policy, legal and regulation) and the contractual arrangements between the principal (Assembly) and the agent (companies) are necessary conditions for solid waste management. The analysis of the regulation in this study revealed that some elements of the current regulatory practices were against the regulatory rules and legislation. Eight elements of weak regulatory practices and non-adherence to formal rules and contractual obligations over the two decade were identified: (1) inappropriate zoning of cities (2) no competitive bidding and weak competition (3) private sector engagement without signed contracts (4) low prices and delay in upward review, (5) delay in payment for service contracts and subsidized franchised service with no interest payment,

(6) weak enforcement of by-laws and sanctions in contracts, (7) undocumented service standards and performance monitoring indicators, and (8) regulatory capture in contract award and paying for services, repeal of a contract clause on interest payment, and review of unit price and charges for service delivery.

The results of regulatory factors influencing private sector performance show that each company operates in a particular delineated zone (no open competition allowed) and some contractual agreements exist between companies and Assemblies. There was a shift from unsigned contract arrangements (giving of letters to notify company to render service and to continue service without signed contract) to providing opportunity for companies to renew contract through competitive bidding and signing of contract. There was non-adherence to contractual obligations by the Assemblies in areas such as review of user charges, payment of services, enforcement of by-law, control of zonal monopoly, payment of interest. There were delays in payment by the Assemblies but no interest paid on monies delayed. The implication of the low prices and delay in payment is that the service quality was compromised. The delay in subsidy payment to the private companies does not provide incentive mechanisms for private sector growth and does not enhance better service delivery.

The Assemblies (municipal authorities) regulate service charges with or without subsidy for franchise house-to-house service and the unit price per tonne paid to the companies for communal collection is set by the Assemblies. The review of the service charges and unit price is delayed for two or more years. There was power relation problem in the review of charges. The implication is that the Assemblies control and allow the charges to be eroded by inflation. The implication of current regulation of prices by the local government is that investment in new collection vehicles cannot be recovered and the service quality of the companies is undermined, since service charges and collection fees are not sufficient to cover full cost of service. The lack of periodic adjustment of the price makes service providers to reduce quality of service to maintain profits. The maintenance of service quality requires monitoring and incentive system (regular price review) to ensure that private service providers do not compromise service quality (OECD, 2000).

The theoretical implication of the non-adherence to contractual obligations by the Assemblies is that there was power relation imbalance and 'regulatory capture' by politicians. Politicians have rent seeking and opportunistic behaviour with 'divide and rule' attitude (Batlay, 2001). The company that received regular payments had political connections with the politicians in government at the time. The regular payment to this company could be attributed to 'regulatory capture' by interest groups in government with rent-seeking and collusion behaviour.

The obligations in contracts are not strictly followed and sanctioning of private companies for non-compliance was not enforced. Both 'relational' and 'competitive' contracting existed in the Ghana solid waste sector, and literature on contracting support this finding (Boyne, 1998). Though certain initial contracts may be awarded on competitive basis, the relationship between the local government and the service provider may be subsequently maintained through negotiation rather than competition for contract renewal. The obligations of the companies in the contracts are, in practice, not strictly adhered to. The companies have reduced house-to-house collection frequency from two to one, and the repairs of communal containers are not done. Trust, cooperation and decrease in the quality of service appeared to be the driving factors for sustaining the private sector in business, but the companies were of the view that they were highly responsive to the Assemblies' requirements of service quality.

Figure 9.2 provides a typology of regulatory practices and adherence to obligation (weak or strong) and service quality (high or low) for solid waste collection. The current regulatory practice is I (for communal service) where cost recovery is low and payment of subsidy is delayed, but there is a shift to IV where there is competitive bidding, contracts are signed and all service recipients are required to pay to improve service quality. The current practice for house-to-house service is III and is shifting to IV since the Assemblies have now combined both services into one contract with more focus on cost recovery through user charging, contract monitoring, and enforcement of sanctions.

The Assemblies had weaknesses in the regulation of solid waste services and defaulted in many of their obligations. The finding revealed Assemblies' non-adherence to obligations, weak financial autonomy, and
weak regulatory capacity. The implication of the regulatory practices is that
the service provision is inefficiently and ineffectively delivered due to low
cost recovery, inadequate monitoring of quality of service, and enforcement of by-laws and sanctions (Figure 9.2). In such a situation, regulation
of prices by the local government, where service charges and collection
fees are not sufficient to cover full cost of service, undermines the service
quality of the companies. The delay in subsidy payment to the private
companies does not provide incentive mechanisms for private sector
growth and does not enhance better service delivery. Weak enforcement of
legislation does not provide sufficient incentives mechanisms for better
service quality and full coverage.

Regulatory practices and adherence to contract obligation

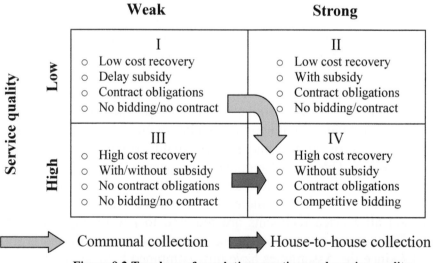

Figure 9.2 Typology of regulation practices and service quality

There was no public company in the solid waste sector to compete with the private sector, and for comparison of efficiency and service quality with the private sector. The use of mixed delivery by private firms and public firms helps create competition in the local markets. It provides local governments with cost of service information and helps to reduce transaction cost and ensure local government has capacity to assume responsibility for service delivery in case of contract failure. The provision of solid waste service by public units under statutes as in the case of 3 cities (Tema, Tamale, and Takoradi) does not provide sufficient information on cost of service information due to hidden costs.

Private sector capacity

The fourth hypothesis for the study states that scale of operation of private companies and the capabilities for solid waste collection influence private sector performance (productivity and service quality). Most of the companies had low productivity due to diseconomies of scale (sub-optimal scale of operation of the companies) and inadequate physical capacity problems. Some reasons for the low productivity and efficiencies are also attributed to inefficient operations and low utilization of vehicles. In theory, the optimal scale for any company is at the highest productivity (efficiency), as this maximises profits and, given perfect competition, this will be the companies' optimum scale, although it may slack away from the optimum from time to time due to effects of other organisational factors. The differences in productivity can be explained by the differences in the inputs and structures of companies (managerial control). The inefficiency may be due to internal structural factor (scale, managerial/employee slack) and external factors (degree of competition, public control) (Mayes et al., 1994). Technical inefficiency is the extent to which companies produce less than they could with their existing allocation of resources. In theory, inefficiency of companies arise from factors such as technology, quality of inputs, scale of operation, allocation of resources and managerial/technical capability (how the company is run) (Mayes et al., 1994).

The theoretical implication of the finding is that productivity of private companies increases with increasing scales of operation and decreases beyond optimum scale of operation. A company's scale of operation or size is often a factor in determining efficiency. The optimum scale of operation (production) is where costs are minimized (at the optimum efficient scale). Any departure from this optimum efficient scale results in scale inefficiency. From theoretical analysis of efficiency, it is possible for a company to be both allocatively and technically efficient, but may be scale inefficient if operation is at a scale either too large or too small to minimise costs. In practice, this tends to occur if a company is too small (suboptimal scale) to be able to utilise its resources efficiently. This was the case for some of the poor performing companies. In contract, the over emphasis on competition compromises the benefits of economies of scale. Again, the benefit of competition is not attained if there is lack of alternative competitive providers and the economies of scale are not taken into consideration (Warner and Bel, 2008). The theory of economies of scale suggests that efficiency is influenced by the level of service output (Boyne, 1998).

The results show that the scale of operation of private companies and the capabilities for solid waste collection supervision influence private sector performance (productivity and service quality). In theory, the inefficient use of resources account for the low productivity and poor service quality of the companies. The implication is that adequate private sector capacity for service provision ensures that resources are used more efficiently and also stimulates innovations to take competitive advantage to achieve high productivity and provide better service quality. In organisational theory, the physical capacity and capability (human resources) of an organization affects or shapes the organization's performance (Lusthaus et al., 2002; Grant, 1996). Management in private organisations has relationship with performance of the companies (Boyne, 2003). The private sector extra resources coupled with prudent management must lead to improved performance in order to maximize profit. Financial resources allow real resources to be purchased and this in turn influences service performance. In theory, the relationship between resources (financial and real resources) and service improvement is positive (Boyne, 2003; Newbert, 2007).

9.5.2 Areas for Further Research

In contract theory, the contract performance monitoring is an integral part of contractual relationship. The theoretical implication is that for the service provider to be monitored, service standards and performance indicators are necessary. The results of this study show that performance and service standards are not well defined in the contract documents. This weakens performance monitoring effort and reporting and there is no systematic approach for that. Further research to establish clearly defined standards and indicators of service quality is needed. This will help in the contracting-out and regulating the private sector activity

The regulation of individual behaviour and the enforcement, in theory, are necessary for public service delivery in the face of market failures (such as externalities and free riding). The by-law enforcement unit and those responsible for contract management are not working together in a coherent manner. Further research to identify the bottlenecks in the enforcement to prevent indiscriminate dumping and to enforce compliance to by-laws and charging for solid waste collection service will provide further understanding. Research into the institutional structure and the legal framework to remove duplication of functions and to enhance maximum impact on solid waste management is warranted.

Further studies on ability to pay and levels of cross-subsidy for urban poor communities need to be conducted. Regulation of user charges need to cover full cost and cross-subsidy for the urban poor, but the current level of service charges seems not to cover full cost of collection and disposal. Detailed cost and service charge analysis need to be conducted to ensure that the private sector companies recover cost of new collection trucks.

Further capacity needs assessment and capacity building plan be developed for the private sector. The study suggests that the private sector needs capacity building on productivity and efficiency analysis to improve solid waste collection.

9.5.3   Implications for practice and policy formulation

The study concludes with policy implications for the design and implementation of private sector involvement in solid waste collection and management. First, to better facilitate and regulate PSI in solid waste collection, clearly defined standards and service quality need to be included in the contract for regulating the private sector activity. This requires capacity building of the officials of Waste Management Department.

Second, the regulation of user charges needs to cover full cost and cross-subsidy for the urban poor to ensure financial sustainability and better service quality. This can be achieved through pay-as-you throw (PAYT) mechanism for communal collection. This mechanism even though it failed in Accra in 1995, it is working well in Kumasi. The success of PAYT depends on public awareness, involvement of the households, and enforcement of by-laws on illegal/indiscriminate dumping. The current fee-based house-to-house collection in poor areas in Accra and monthly payment of service charges by poor households is not sustainable. The implication is that poor people are served with higher service level (house-to-house collection) which is costly compared to communal collection. PAYT affords the households the opportunity to reduce cost through waste reduction and recycling, and it will cost them half of what they are paying now for the monthly fixed charges.

Third, the by-law on indiscriminate dumping should be enforced alongside charging for solid waste collection service. Most of the respondents were aware of the measures to prevent indiscriminate dumping, but private sector and Assemblies are not responsive to service delivery. There is the need to restructure the environmental health unit to make it more responsive to the challenges of sanitation and solid waste collection and management. The health inspectors from the environmental health unit can be seconded to the private companies for them to function more effectively as officers from the Assemblies working with companies to enforce by-laws on paying for the service and prevention of illegal/indiscriminate dumping of waste.

Fourth, the results on waste minimisation efforts showed that there is no formal arrangement for waste reduction and recycle. There is household willingness to source separate organic waste and willing-to-pay for service. There is the need for formalising waste minimisation to divert greater quantity of waste from the disposal stream.

Fifth, the current solid waste collection situation needs improvement through capacity building of the technical operations and commercial staff of the private sector companies. The capacity of Assemblies needs improvement in order to be better able to facilitate, monitor, and regulate the private sector.

REFERENCES

Abgemabiawo, l., K., (1996). Socio-economic and institutional dimensions of solid waste management in Tema, Ghana. MSc thesis, Norwegian University of Science and Technology, Trondheim.

Adarkwa K. K. (2005). The reality of decentralisation implementation in Ghana and some emerging concerns. In Egziabher T. G. and Dijk M. P. van (eds). Issues and challenges in local and regional development, Addis Ababa,

AMA/Colan Consult (1995). Public and private sector management of solid waste in Accra: An evaluation and prospects for sustainable service delivery.

Aucoin P. (1990). Administration reform in public management: paradigms, principles, paradoxes and pendulums, *International journal of policy and administration* Vol **3** (2) pp115 –137

Archer, K. (1987). The polities of privatization. London. Macmillian.

Awortwi N. (2002). Managing multiple modalities of delivering basic services in Ghanaian cities. *African development* Vol XXVII (3) pp76 – 105.

Awortwi N. (2003). Getting the fundamentals wrong: Governance of multiple modalities of basic services delivery in three Ghanaian Cities. PhD Thesis. ISS.

Awortwi N. (2004). Getting the fundamentals wrong: woes of public-private partnerships in solid waste collection in three Ghanaian cities. *Public administration dev.* Vol **24** pp213–224

Baud, I. (2004). Markets, Partnerships and Sustainable Development in Solid Waste Management: raising the questions. Solid Waste Management and Recycling; Actors, Partnerships and Policies in Hydrebad, India and Nairobi, Kenya.

Baud, I., Post, J. and Furedy, C. Dordrecht/Boston/London, Kluwer Academic Publishers. **76:** 1-20.

Baud, I. and J. Post (2003). Between Market and Partnerships: Urban Solid Waste Management and Contribution to Sustainable Development. GBER **3**(1): 46 - 65.

Baud, I., J. Post and C. Furedy (2004). Solid Waste Management and Recycling; Actors, Partnerships and Policies in Hydrebad, India and Nairobi, Kenya. Dordrecht/Boston /London, Kluwer Academic Publishers.

Ballance T., and Taylor A., (2005). *Competition and Economic Regulation in Water: The Future* of European water industry. IWA Publishing, London, UK.

Baldwin and Cave (1999). *Understanding Regulation,* Oxford, Oxford University Press, pp. 76-85.

Bartone, C.R., L. Leite, T. Triche, and R. Schertenleib (1991). Private Sector Participation in Municipal Solid Waste Service: Experiences in Latin America.Waste Management and Research, 9(6):495-509

Bartone C., R., (2001). The role of the private sector in municipal service delivery in developing countries: keys to success: In Freire, M., and Stren, R., (ed.). The challenges of urban government: policies and practices. WBI Development Series, Centre for Urban and Community Studies, University of Toronto, Washington, DC 2001, pp. 214-223.

Batley R., (2001). Public private partnership for urban services: In Freire, M., and Stren, R., (ed.). The challenges of urban government: policies and practices

Batley R. and Larbi G., (2004) The changing role of government. The reform of public services in developing countries. Palgrave Macmillan, New York.

Batley, R. (1996) Public-Private Relationships and Performance in Service Delivery, *Urban Studies* **33** (4-5), pp. 723–751.

Boyne G. A., (2003). Sources of public service improvement: A critical review and research agenda. Journal of public administration research and theory. Vol 13, no 3 pp 367–394.

Boston J. (2000). The challenge of evaluating systemic change: the case of public management reform. *International public management journal.* **3** pp 23–46.

Boakye, C. (2005). Solid Waste Collection Systems in Ghana. Urban Forum on Solid Waste Primers and Lessons Learned from World Bank Projects. March 7, 2005, Washington D.C.

Bovaird, T., (2006). Developing new forms of partnership with the 'market' in the procurement of public services. *Public Administration* vol. **84**, no. 1, 2006 (81–102)

Benneh et al., (1993). Environmental problems and urban household in the Accra Metropolitan Area. Stockholm Environment Institute.

Bel, G., and Warner M., (2008). Does privatization of solid waste and water services reduce costs? A review of empirical studies. *Resources, Conservation and Recycling,* 52 (2008) 1337–1348

Bel G, and Costas A., (2006). Do public sector reforms get rusty? Local privatization in Spain. *Journal of Policy Reform* 9 (1):1–24.

Brown T., L., and Potoski M., (2004). Managing the public services market. Public Administration 64 (6), 656-668,

Boyne G., A., (1998). Bureaucratic theory meets reality: public choice and service contracting in UN local government. Public Administration 58(6), 474 – 484

Carl Bro International/Colan Consult (1998). Private participation in solid waste management. Ministry of local government, Ghana,

Chang H., (1997). The economics and politics of regulation Cambridge Journal of Economics, 21, 703-728

Carter N., Klein R., and Day P., (1992). How organisations measure succuss. The use of performance indicators in government. Routledge, London.

Carmines, E. G. and R. A. Zeller (1994). Reliability and validity assessment. Basic Measurement. M. S. Lewis-Beck. Thousand Oaks, CA, Sage: 1-58.

Coelli, T.J., Rao, D.S.P., O'Donnell, C.J. and G.E. Battese (2005) An Introduction to Efficiency and Productivity Analysis, 2nd Edition. Springer, New York

CEP (2008). DEAP software developed by Centre for Efficiency and Productivity Analysis (CEP), School of Economics, University of Queensland (UQ)

Cointreau Levine (2005).Waste collection system in developing countries. Seminar on Solid Waste Primers and Lessons Learned from World Bank Projects. March 7, 2005, Washington.

Cointreau-levine S (1994). Private Sector Participation in Municipal Solid Waste Management, Urban Management Programme Paper No 13, World Bank, [http://www.developmentbookshop.com/detail-aspx?ID=1039], (accessed 2006 February 28).

Cointreau-Levine, and Adrian Coad (2000) Guidance Pack: "Private Sector Participation in Municipal Solid Waste Management," [http://www.sanicon.net/titles/title.php3?titleno=472], (accessed 2006 February 28).

Cointreau-levine S (1999). Private Sector Participation in Municipal Solid Waste Management. In UNDP Public-private partnerships for the urban environment. UNDP/Yale.

Cook, P. and Kirk-Patrick, C. (1995). Privatisation policy and performance: international perspectives. London. Prentice-Hall.

Cooper W. W., Seiford L., M., And Tone K., (2000). Data Envelopment Analysis. A Comprehensive Text with Models, Applications, References and DEA-Solver Software, Khmer Academic Publishers, Boston/Dordrecht/London

Cronbach, L. J. (1951). "Coefficient alpha and the internal structure of tests." Psychometrika 6: 297-334.

Cullivan, D, Tippett, B., Edwards, D., B., Rosensweig, F., and McCaffery J.,(1988). Guidelines for Institutional Assessment Water And Wastewater Institutions, Wash Technical Report No 37

Callan S., J, and Thomas J., M., (2001). Economies of scale and scope: a cost analysis of municipal solid waste services. *Land Economics,* 77(3):548–60.

Collins J., N, and Downes B., T., (1977). The effect of size on provision of public services: the case of solid waste collection in smaller cities. *Urban Affairs Quarterly, 12* (3):333–47.

Downs, A. (1967), Inside Bureaucracy, Rand Corporation Research Study, Little, Brown, and Company: Boston.

Deakin N. and Walsh K. (1996). The enabling state: the role of markets and controls. *Public administration* Vol **74** pp33–48

Domberger, S. and Hall C. (1996). Contracting for public services: A review of Antipodean Experience. *Public administration* Vol **74** pp129–47

Domberger, S., Meadowcroft, S. and Thompson D. (1986). Competitive tendering and efficiency: the case of refuse collection. *Fiscal studies* 7(4): 69-87

Dunleavy, P. and Hood, C. (1994). From old public administration to new management. *Public money and management* 14 (3). 9–16.

Dijkgraaf E, and Gradus R.,H.,J., M. (2003). Cost savings of contracting out refuse collection. *Empirica*, 30(2):149–61.

Dijkgraaf E., and Gradus R.,H.,J.,M. (2007). Collusion in the Dutch waste collection market. *Local Government Studies*, 33(4):573–88.

Dijkgraaf E., and Gradus R.,H.,J.,M. (2008). Institutional developments in the Dutch waste collection market. *Government and Policy*. **26**(1):110–26.

Donahue J.,D. (1989). The privatization decision. Public ends, private means. New York, NY, USA: Basic Books,

Dubin J.,A, and Navarro P. (1988). How markets for impure public goods organize: the case of household refuse collection. *Journal of Law, Economics & Organization*, 4(2):217–41.

Deaken N., and Walsh K., (1996). The enabling state: the role of markets and contracts. Public Administration, 74, 33-48

Etzioni A., (2009). The capture theory of regulations—revisited, Soc 46:319–323,

Eggert (2005). Evolving face of solid waste management, waste management and research,

Folz, D., H., and Lyons, W., (1986).The Measurement of Municipal Service Quality and Productivity: A Comparative Perspective Public Productivity Review, Vol. 10, No. 2, pp. 21-33

Franceys, R., and Gerlash, E., (2008). *Regulating water and sanitation for the poor.* Economic regulation for public and private partnerships. Earthscan UK,

Field B. C., (1997). Environmental economics. An introduction, 2 ed. McGraw-Hill.

Fernandez, S., (2005). Accounting for Performance in Contracting for Services: Are Successful Contractual Relationships Controlled or Managed? Paper presented at the 8th Public Management Research Conference, Los Angeles, California, and hosted by the School of Policy, Planning, and Development at the University of Southern California, September 29 to October 1, 2005.

Gafishi M., C., (2009). Public perception on solid waste collection services, case study of Kigali city of Rwanda. MSc Thesis, KNUST,

Gómez-Lobo A, and Szymanski S. A. (2001). Law of large numbers: bidding and compulsory tendering for refuse collection contracts. *Review of Industrial Organization*, 18(1):105–13.

Gerlash, E., (2008). Regulating for poor. In: Franceys, R., and Gerlash, E., (2008). Regulating water and sanitation for the poor. Economic regulation for public and private partnerships. Earthscan UK,

Gidman, P., Blore, I., Lorentzen, J. and Schuttenbelt, P., (1999). Public-private partnership in urban infrastructure services, UPM working paper series 4, 1-11, UNDP/Habitat/World Bank, Nairobi.

Grimshaw, Vincent and Willmott, (2002). *Going privately: partnerships and outsourcing in UK public service. Public administration.* **80** (3):475 – 502

Ghana Statistical Service (2002). Population and housing census 2000.

Grand J., (1991). The Theory of Government Failure, *British Journal of Political Science*, Vol. 21, No. 4, pp. 423-442

Grant, R. M. (1991). The resource-based theory of competitive advantage: Implications for strategy formulation. California Management Review, 33(3), 114–135.

Grant, T., (1996). Keys to successful public private partnership, Canadian Business review, 23 (3), 27- 28,

Hardcastle, C., and Boothroyd, K., (2003). Risk overview in public private partnership, In: Akintoye, A., Bek, M., and Hardcastle C., (2003), PPPs, managing risks and opportunities, Blackwell Science, Oxford, UK.

Hansen, G. S., & Wernerfelt, B. (1989). Determinants of firm performance: The relative importance of economic and organizational factors. Strategic Management. Journal, 10, 399–411

Hirsch W., Z. (1965). Cost functions of an urban government service: refuse collection. *Review of Economics and Statistics*, 47(1):87–92.

Hantke-Domas M., (2003). The public interest theory of regulation: nonexistence or misinterpretation. *European journal of law and economics*, **15**:165-194.

Helmsing A.H.J., (2002). Decentralization, enablement and local government in low-income countries. *Government and policy* **20**:317-40

Helmsing A.H.J., (1997). Local government and finance. An introduction, ISS, The Hague.

Hood C. (1991). A public management for all seasons. *Public administration* Vol **69** pp3–19

Hood C. (1995). Contemporary public management: a new global paradigm. *Public policy and administration* Vol 10 (2)

Heugens, P., P., M., A., R, (2001). Strategic Issues Management: Implications for Corporate Performance. ERIM Ph.D. Series Research in Management 7.

Hoggett P., (1996). New modes of control in the public service. Public Administration. 74, 9-32.

Jamison M., A., and Berg S., V., (2008). Annotated reading list for a body of knowledge on infrastructure regulation.

Kassim S .M. (2006). Sustainability of Private Sector in Solid Waste Collection – A Case of Dar es Salaam Tanzania PhD Thesis, Loughborough University UK, 2006

Kassim S., M., and Ali M., (2006). Solid waste collection by the private sector: Households' perspective – Findings from a study in Dar es Salaam city, Tanzania, *Habitat International* 30: 769–780

Karikari, O., (1999). Community participation and choice of technology in municipal solid waste management: Case study of Accra. MSc thesis, Vrije Universiteit, Brussels.

Karanja A., (2005). Solid waste management in Nairobi: actors, institutional arrangements and contribution to sustainable development. PhD Dissertation, ISS.

Kessey K. D. (1995). Financing local development in Ghana. mobilization and management of fiscal resources in Kumasi Metropolis. PhD Dissertation.

Kessides, C. (1993). Institutional options for the provision of infrastructure. World Bank, Washington D. C.,

Eggerth, L., L., (2005). The evolving face of private sector participation in solid waste management, *Waste Management*, 25 (3), p.229,

Kidder, L. and C. M. Judd (1986). Research methods in social relations. New York, Holt, Rhinehart & Winston.

Kemper P, and Quigley J. (1976). The economics of refuse collection. Cambridge, MA, USA: Ballinger; 1976.

Kirkpatrick C, Parker D, Zhang Y., F. (2006). State versus private sector provision of water services in Africa. *World Bank Economic Review*, 20(1):143–63.

Kitchen H., M. (1976). A statistical estimation of an operating cost function for municipal refuse collection. *Public Finance Quarterly,* 4(1):56–76.

Kite News (2008). Solid waste management situation in Kumasi.

Mbuligwe S. E., (2004). Assessment of performance of solid waste management contractors: a simple techno-social model and its application. *Waste management* **24** pp 739–749

Ministry of Local Government and Rural Development (MLGRD) (1999). Environmental Sanitation Policy. Accra: Government of Ghana. P 3, 17-20.

Massoud M., A, El-Fadel M, Abdel Malak A. (2003). Assessment of public vs. Private MSW management: a case study. Journal of Environmental Management, 69(1):15–24.

Minogue, M., (1997). Theory and Practice in Public Policy & Administration. In Hill (Ed) The Policy Process; A Reader. pp10-29.

Mwangi, S. W. (2003). Challenges of urban environmental governance. Participation and partnerships in NuKuru Municipality, Kenya. PhD Thesis Amsterdam University.

METAP (2009). Municipal solid waste management. [Accessed 2009 from www.metap-solidwaste.org, METAP SWM Centre, guidelines of PSP, World Bank and GTZ project,

Mayes, D., Harris C., and Lansbury M., (1994). Inefficiency in Industry, Harvester, London

Newbert S., L., (2007) empirical research on the resource-based View of the firm: an assessment and Suggestions for future research, *Strategic Management Journal*, 28: 121–146

Nachum, L., (1999). The origin of the international competitiveness of firms. Impact of location and ownership in professional service industries. Edward Elgar, Cheltenham.

Niskanen, W. (1994), Bureaucracy and Public Economics, Cheltenham: Edward Elgar Publishing Limited.

Lusthaus C., Anderson, G., and Murphy, E., (1995). Institutional assessment, A Framework for Strengthening Organizational Capacity for IDRC's Research Partners,

Lusthaus, C. Adrien, M., H., Anderson, G., and Carden, F., (1999). Enhancing Organizational Performance. A Toolbox For Self-Assessment. IDRC/IDB, International Development Research Centre.

Lusthaus, C. Adrien, M., H., Anderson, G., Carden, F., and Montalvan, G., P., (2002). Organizational assessment. A framework for improving performance. International Development Research Centre.

Laryea-Adjei, G., Q., M., (2006). Central local relations in the provision of basic services, provision of water and sanitation servives in Ghana. PhD thesis, Erasmus University.

Loehr, W. and Sandler, T. (1998). On the public character of goods. In Loehr and Sandler (eds) Public goods and public policy. Beverly Hills, London. Sage Publications.

Li, B., and Akintoye, A., (2003). An overview of public private partnership, In: Akintoye, A., Bek, M., and Hardcastle C., (2003) (ed), PPPs, managing risks and opportunities, Blackwell Science, Oxford, UK.

Obeng P., (2005). Assessment of Institutional Arrangements for Solid Waste Management in Kumasi., KNUST MSc Thesis,

Obirih-Opareh, N. (2002). Solid waste collection in Accra. The impact of decentralisation and privatisation on the performance of service delivery. PhD dissertation.

Obirih-Opareh, N., Eran Razin, Sjaak van der Geest and Johan Post (2004). The effects of decentralisation and privatisation on urban environmental management: waste management in the Accra Metropolitan Area. NIRP Research for Policy Series 22

Obirih-Opareh N., and Post J., (2002). Quality assessment of public and private modes of solid waste collection in Accra, Ghana. *Habitat International 26 pp 95–112.*

Oduro-Kwarteng S., Awuah E., and Kotoka P., (2006). Solid waste management practices in Kumasi, Ghana: public concerns and attitude towards disposal and cost recovery.

Oduro-Kwarteng, S. and Dijk, M. P. V., (2008). Performance of private companies involved in urban solid waste management: Evidence from three cities in Ghana. *Proceedings of the 33rd WEDC International conference on water and sanitation, 7-11th April 2008, Accra.*

Oduro-Kwarteng S. (2009). Institutional arrangements for private sector involvement in urban solid waste collection: Case study of five cities in Ghana. *Proceedings of 34th WEDC International Conference on Water, sanitation and hygiene: sustainable development and multi-sectoral approaches, Addis Ababa, Ethiopia, 11-22 May 2009,*

Obsborne D. and Gaebler T. (1993). Reinventing government. How the entrepreneurial spirit is transforming the public sector. Penguin Group. USA.

Odle, M. (1993). Towards a stage theory approach to privatization. *Public administration and development* Vol 13, 17–35

Onibokun and Kumuyi, (1999). Governance and Waste Management in Africa: In Adepoju G. and Onibokun (ed). Managing the monster urban waste and governance in Africa. Canada's International Development Research Centre.

Otieno, J.T. and Shem, O.W, (2000). Sustainable municipal solid waste management and improvement of social services delivery. *Regional Development Dialogue* 21(2).

Organisation for Economic Cooperation and Development (1993). Managing with market-type mechanisms, Paris, OECD.

Ohlsson H. (2003). Ownership and production costs. Choosing between public production and contracting-out in the case of Swedish refuse collection. *Fiscal Studies,* 24(4):451–76.

OECD (2000). Competition in Local Services: Solid Waste Management, OECD Directorate for Financial, Fiscal and Enterprise Affairs Committee on Competition Law and Policy

Owusu-Ansah, K. (2008). Evaluation Of Household Solid Waste Processing Options In Accra, Ghana. Msc Thesis, UNESCO-IHE, Delft.

Ulkuniemi, P., (2003). Purchasing Software Components At The Dawn Of Market, Thesis, Department of Marketing, University of Oulu. Oulu,

Pollitt, C. (1990). Managerialism and the Public Service: the Anglo-American Experience. Oxford: Basil Blackwell

Post J., (1999). The problems and potential of privatizing solid waste management in Kumasi, Ghana. *Habitat International 23 (20) pp201–215.*

Post J., Broekema J. and Obirih-Opareh N. (2003). Trial and Error in Privatisation: Experiences in Urban Solid Waste Collection in Accra (Ghana) and Hyderabad (India). *Urban studies,* **40** (4): 837-854

Pommerehne W.W, and Frey B. (1977). Public versus private production efficiency in Switzerland: a theoretical and empirical comparison. *Urban Affairs Annual Review,* 12:221–41.

Prager, J. (1994). Contracting-out Government services: lessens from the private sector. *Public administration review.* **54** (2):176–84

Prasad N., (2006). Privatisation Results: Private Sector Participation in Water Services after 15 Years. *Development Policy Review,* 2006, 24 (6): 669-692

Pier W.J, Vernon R.B, and Wicks J.H. (1974). An empirical comparison of government and private production efficiency. *National Tax Journal*, 27(4):653–6.

Reeves E, and Barrow M. (2000). The impact of contracting-out on the costs of refuse collection services. The case of Ireland. *Economic and Social Review*, 31(2):129–150.

Roth, G. (1987). The private provision of public services in developing countries. Oxford University Press, New York.

Rothenberger, D., Frei, U., and Brugger, F., (2005). Policy principles and implementation guidelines for private sector participation in the water sector – a step towards better results. *Water Science & Technology* Vol 51 No 8 pp 61–69 Q 2005

Rotich K. H., Yongsheng Z., and Jun D., (2006). Municipal solid waste management challenges in developing countries – Kenyan case study, Waste Management 26: 92–100

Rakodi C. (2003). Beyond public failure and private success: disentangling theory, practice and outcomes in the provision of urban environmental services. Paper presented at the N-AERUS conference, Paris (15–17 May).

Rondinelli (1999). Decentralisation briefing notes, WBI working papers, WBI/PREM Network, Washington.

Savas E. S. (1978). On Equity in Providing Public Services, Management Science, Vol. 24, No. 8, pp. 800-808

Saade, L., M.W. Blokland and Brikke (2000): Vision 21. Document by the Water Supply and Sanatition Collaborative Council.

Schubeler, P., Wehrle, K., and Christen, J. (1996). *Conceptual Framework for Municipal Solid Waste Management in Low-Income Countries.* Swiss Centre for Development Cooperation in Technology and Management, St. Gallen.

Schwartz K., (2006a). Water services management, lecture notes, UNESCO-IHE, Delft.

Schwartz K., (2006b). Managing public water utilities: An assessment of bureaucratic and new public management models in the water supply and sanitation sectors in low- and middle-income countries. PhD thesis. Erasmus University, Rotterdam.

Schouten M., (2009). Strategy and performance of water supply and sanitation providers, effects of two decades of neo-liberalism. PhD Thesis, Erasmus Universiteit Rotterdam, The Netherlands,

Sullivan, J. L. and S. Feldman (1994). Multiple indicators: An introduction. *Basic Measurement.* M. S. Lewis-Beck. Thousand Oaks, CA, Sage: 59-137.

Stren, R., (2001). Private involvement in provision of public service: In Freire, M., and Stren, R., (ed.). The challenges of urban government: policies and practices

Stevens B.J. (1978). Scale, market structure, and the cost of refuse collection. *Review of Economics and Statistics,* 60(3):438–48.

Szymanski S. (1996). The Impact of compulsory competitive tendering on refuse collection services. *Fiscal Studies*, 17(3):1–19.

Szymanski S, and Wilkins S. (1993). Cheap rubbish? Competitive tendering and contracting out in refuse collection. *Fiscal Studies*, 14(3):109–30.

Saleh J., A., (2002). Assessement of a pilot franchised solid waste collection scheme in Kumasi. MSc thesis, KNUST, Kumasi.

Skumatz L., (2002). Variable-Rate or Pay-as-you-throw Waste Management, Reason Foundation,

Teske, P., (2004). Regulation in the states, Brookings Institution, Washington, DC:

Tchobanoglous, G., Theisen, H., and Vigil, S. (1993). *Integrated waste management: Engineering principles and management issues,* McGraw-Hill, New York.

Tsiboe, I., A., and Marbell, E., (2004). '*A look at urban waste disposal problem in Accra, Ghana*' MSc thesis, Roskilde University

Tickner G, and McDavid J. C. (1986). Effects of scale and market structure on the costs of residential solid waste collection in Canadian cities. Public Finance Quarterly, 14(4):371–93.

Tangen S., (2005). Demystifying productivity and performance. International Journal of Productivity and Performance Management Vol. 54 No. 1, 2005 pp. 34-46

Tullock, G. (1965), The politics of bureaucracy. Washington DC. Public Affairs Press.

UNIDO/UNDP (1998). Manual on municipal solid waste management for Africa,

UNDP (2008). Capacity assessment, practical note. United Nations Development Programme

United Nations (1993). The Global Partnership for Environment and Development: A Guide to Agenda 21. New York, United Nations: 88 - 94.

United Nations Habitat (1996). An Urbanizing World, Global Report on Human Settlements 1996. New York, United Nations Centre for Human Settlements (HABITAT):

van Dijk, M., P., (2004a). Institutional Options For Wastewater and Sanitation Reform, World Water Week in Stockholm on Water resources, decision-making processes and principles, A contribution to Panel 3 (17-8-4),

van Dijk M. P. Van (2004b). Research methods and tools for an MBA course in Nanjing, Nanjing University.

van Dijk M. P. Van (2006). Managing cities in developing countries. theory and practice of urban management. Edward Elgar, UK.

van Dijk, M. P. V., and Oduro-Kwarteng, S. (2007). Urban management and solid waste issues in Africa. *Proceedings of the International conference on solid waste and climate change, 23rd-27th September 2007, International Solid Waste Association, Amsterdam.*

van Dijk, M., P., V., (2008a). Private sector involvement in water and sanitation, *International Journal of Water*, Vol. 4, No.3/4

van Dijk, M., P., V., (2008b). Public–private partnerships in basic service delivery: impact on the poor, examples from the water sector in India, *International Journal of Water* 2008 - Vol. 4, No.3/4, pp. 216 - 234

van de Klundert A., and Anschütz J., (2001). Integrated Sustainable Waste Management - the Concept. Tools for Decision-makers Experiences from the Urban Waste Expertise Programme (1995-2001)

WATER-AID GHANA/WEDC (2005). Assessment of National Sanitation Policies -Ghana Case, Final Report, Application of tools to support sanitation Policies.

Walsh, K. (1995). Public services and market mechanisms: competition, contracting and the new public management. Machmillan Press Ltd, London.

Warner M., E., and Bel G., (2008). Competition or monopoly. Comparing privatization of local public services in the USA and Spain. Public Administration 86(3), 723-735,

Wilson D., Whiteman A., and Tormin A., (2001). Strategic planning guide for municipal solid waste management. World Bank, DFID,

Wilson, D., C., (2007). Development drivers for waste managementWaste Management & Research, 25: 198–207

World Bank (1994). World development report 1994: Infrastructure for development. Oxford and New York, Oxford University Press.

World Bank (1997). World development report 1997: The state in the changing world. Washington, D.C:World Bank.

World Bank (2004). World development report 2004: Making services work for the poor. Washington, D.C:World Bank.

WCED (1987). Our Common Future. World Commission on Environment and Development, Oxford University Press, Oxford.

WEDC (2003). Application of tools to support national sanitation policies. Water, Engineering and Development Centre,

Zurbrugg, C., (1999). Solid Waste Management in Developing Countries. SANDEC / EAWAG;

ANNEX

ANNEX 1 RELIABILITY AND UNIDIMENSIONALITY

Private sector capacity

The vehicle operations capability was measured using two variables: vehicle route planning and operations scheduling. Five-item measure of the operation capability was separated into two variables to improve upon the internal consistency of the scale. A five-point scale ranging from 1 (strongly disagree) to 5 (strongly agree) was used to measure each item of the measurement scale. The reliability analysis of the vehicle route planning scale showed that the corrected item-total correlations and the inter-item correlations were high resulting in a Cronbach's alpha of 0.73. The reliability analysis showed a satisfactory degree of internal consistency, since a Cronbach's alpha of 0.60 is satisfactory for exploratory research. Furthermore, a validity analysis to assess unidimensionality of the scale using principal components analysis revealed that the first component had an eigenvalue of 1.97 (on a three-item scale), and consequently explained 65 % of the variance. The second component had an eigenvalue of 0.69, explaining 23 % of the variance. Thus, the route planning scale developed was unidimensional, since the three-item scale has only one component with eigenvalue greater than one with more than 60% of variance explained.

The reliability analysis of the operation scheduling scale showed a satisfactory degree of internal consistency. The inter-item correlation was high with a Cronbach's alpha of 0.64. The first component had an eigenvalue of 1.32 (on a two-item scale), and consequently explained 65 % of the variance. The second component had an eigenvalue of 0.68 explaining 34 % of the variance. Thus the operation scheduling scale developed was unidimensional. Therefore, it may be concluded that the two variables

331

scale for the operation capability developed were unidimensional and reliable.

The operations supervision capacity was measured using two variables: supervisory capacity and employees' moral to work satisfactory. Each of the variables was measured using two items. The four-item measure of the operations supervision capacity was separated into two variables to improve upon the internal consistency of the scale. A five-point scale ranging from 1 (strongly disagree) to 5 (strongly agree) was used to measure each item of the measurement scale. The reliability analysis of the supervisory scale showed that the inter-item correlation was high with a Cronbach's alpha of 0.66. The reliability analysis showed a satisfactory degree of internal consistency. Also, a principal components analysis to analyse validity of the scale revealed that the first component had an eigenvalue of 1.497 (on a two-item scale), and consequently explained 75 % of the variance. The second component had an eigenvalue of 0.50, explaining 25 % of the variance. Thus, the supervisory capacity scale developed was unidimensional.

The reliability analysis of the employee moral scale showed a satisfactory degree of internal consistency. The inter-item correlation was high with a Cronbach's alpha of 0.82. The first component had an eigenvalue of 1.72 (on a two-item scale), and consequently explained 86 % of the variance. The second component had an eigenvalue of 0.28 explaining 14 % of the variance. The employee moral scale developed was unidimensional. It is therefore concluded that the two variables scale for the supervision capacity developed were unidimensional and reliable. The responses of the various companies with regards to vehicles operations supervision.

The maintenance capability was measured using two variables: the availability of in-house capacity for vehicle maintenance and vehicle maintenance schedules. The in-house capacity is a dummy variable, whereas the maintenance schedule variable is a survey item measured using a five-point scale ranging from 1 (strongly disagree) to 5 (strongly agree). The maintenance capacity is one of the factors influencing performance of private solid waste collection companies. Vehicles maintenance under this study was assessed in terms of whether vehicle maintenance schedules were prepared and followed and availability of in-house maintenance workshop and the kind of maintenance (minor or major) that is done.

Regulation

The reliability analysis of the regulation of quality of service shows a satisfactory degree of internal consistency with a Cronbach's alpha of 0.83. A validity analysis using principal components analysis showed that the scale was unidimensional.

The regulation of access to information was measured using three indicators which sought to assess performance targets, access to information on waste quantities and cost of services. The reliability analysis of the regulation of access to information shows a satisfactory degree of internal consistency with a Cronbach's alpha of 0.65. A validity analysis using principal components analysis showed that the scale was unidimensional.

The regulation of user charges was measured using two variables: tariff setting arrangement and level of tariff. The reliability analysis of the regulation of user charge showed a satisfactory degree of internal consistency with Cronbach's alpha of 0.87 and 0.80 for the two variables, tariff setting arrangement and level of tariff respectively. A validity analysis using principal components analysis shows that the scale was unidimensional. The reliability analysis of the regulatory capacity and autonomy show a satisfactory degree of internal consistency with a Cronbach's alpha of 0.75. A validity analysis using principal components analysis showed that the scale was unidimensional.

The regulation of quality of service was measured using four-item scale in Table 7.16. A five-point scale ranging from 1 (very inadequate) to 5 (very adequate) was used to measure each item of the measurement scale. The reliability analysis of the regulation of quality of service scale showed a satisfactory degree of internal consistency. Both the corrected item-total correlations and the inter-item correlations were high with a Cronbach's alpha of 0.83.

Regulatory capacity of the Assembly and Autonomy

The regulatory capacity of the Municipal Assembly was measured using four-item scale in Table 7.19. A five-point scale ranging from 1 (very inadequate) to 5 (very adequate) was used to measure each item of the measurement scale. The reliability analysis of the regulatory capacity scale showed a satisfactory degree of internal consistency. Both the corrected item-total correlations and the inter-item correlations were high with a Cronbach's alpha of 0.75.

The autonomy of the Municipal Assembly to regulate the private sector activities was measured using two variables: financial and managerial autonomy. The financial autonomy of the Municipal was measured using four-item scale about autonomy to set tariff and budget for solid waste management. A five-point scale ranging from 1 (very low) to 5 (very high) was used to measure each item of the measurement scale. The reliability analysis of the financial autonomy scale showed a satisfactory degree of internal consistency. Both the corrected item-total correlations and the inter-item correlations were high with a Cronbach's alpha of 0.81.

The managerial autonomy of the Municipal Assembly was measured using two-item scale about autonomy to terminate contract of private companies and also to enforce by-laws for solid waste management (Table 7.25). A five-point scale ranging from 1 (very low) to 5 (very high) was used to measure each item of the measurement scale. The reliability analysis of the managerial autonomy scale showed a satisfactory degree of internal consistency. Both the corrected item-total correlations and the inter-item correlations were high with a Cronbach's alpha of 0.78.

Assemblies' adherence to contract
The reliability analysis of the contract obligation showed a satisfactory degree of internal consistency with a Cronbach's alpha of 0.86. The first component had an eigenvalue of 2.87, and consequently explained 72 % of the variance. The second component had an eigenvalue of 0.71 explaining 18 % of the variance. Therefore, it may be concluded that the contract obligation scale was unidimensional and reliable.

ANNEX 3 QUESTIONNAIRE

RESIDENTS' SURVEY QUESTIONNAIRE

Household Characteristics

1.1 City ……………………………… Locality/Suburb …………………
1.2 Respondent Age-…………………… Sex : M/F………………………………..
1.3 Is respondent household head?. □Yes …□no…..
1.4 Educational level of Respondent – a) None   b) Basic c)Secondary/tech.
     d)Tertiary
1.5 Occupation of the household head (s)
     (1) Unemployed   (2) Trading   (3) Government employee/Retired
     (4) Self-employed   (5) Agriculture/farming   (6) Others, specify……………
1.6 How many households/families are in the house?…………………………
1.7 How many are you in the house (population of all households)?……………………
1.8 Do you rent the house? Yes….. No……..Rent per month………

Public Awareness and appreciation of waste collection service quality

2.1 Where do you empty or dispose your waste?
     a. Door-to-door collection ………......................................…..........
     b. Communal collection (a skip container in the neighborhood)…………........
     c. An open dump (waste is collected regularly)………………........
2.2 Give name of service provider collecting your waste…………………………
2.3 Has the Assembly or the service provider informed or educated you about your re-
     sponsibility and that of the service provider? Yes….. No……………
2.4 What have you been inform about?………………………………
2.5 Are you happy or satisfy about the quality of collection service? □Yes …□no…..
2.6 Why are you satisfied (or not satisfied) about the quality of collection ser-
     vice?……………………………
2.7 Does the waste in container overflows before container is collected? □Yes …□no…..
2.8 Why does the bin/container overflows?…………………………………………
2.9 How often does the vehicle comes for collection?
          a)   every day,   b) once a week   c) twice a week   d) thrice a week

2.10 How will you rate the quality of waste collection service in the city *using these*
     Scales (1) very poor (2) poor (3) fair (4) good (5) very good

	1	2	3	4	5
(a). Reliability of waste collection (timely collection)	□	□	□	□	□

(b). Sanitary conditions at bin/container site and spill-over of waste ☐ ☐ ☐ ☐ ☐

(c). Quick response to residents' complaints ☐ ☐ ☐ ☐ ☐

(d). cleanliness of surroundings (rating of 5 for no littering) ☐ ☐ ☐ ☐ ☐

(e). Sanitary conditions of drains in the area ☐ ☐ ☐ ☐ ☐

(f). Neatness of waste collection crew, wearing of protective clothing ☐ ☐ ☐ ☐ ☐

(g). Neatness of waste collection vehicles ☐ ☐ ☐ ☐ ☐

(h). Behaviour/attitude of collection crew towards residents ☐ ☐ ☐ ☐ ☐

2.11 How do you rate the effects/impacts of solid waste in your community?
Use these Scales (1) high negative effect (2) negative effect (3) no effect (4) positive effect (5) high positive effect

		1	2	3	4	5
(a).	Impact of waste uncollected on human health,	☐	☐	☐	☐	☐
(b).	Impact of waste uncollected in the neighborhood on aesthetics	☐	☐	☐	☐	☐
(c).	Impact of offensive odour in the neighborhood on health,	☐	☐	☐	☐	☐
(d).	Impact of choke drains with waste/breeding of mosquitoes	☐	☐	☐	☐	☐

2.12 What should be done to prevent wastes being dumped in drains and other unauthorized places?…………………………………………………………………………….

2.13 What illness in your house do you associate with the sanitation conditions in your area?...

2.14 Are you involved in service quality monitoring?...

Waste reuse, recycle and waste minimisation

2.15 What do you do with food waste you generate?...

2.16 What do you do with plastic waste you generate?...

2.17 What waste do you sell?..

2.18 What waste do you give out as gifts?..

2.19 How do you minimize waste being generated?...

2.20 Would you like to separate your food waste and plastic waste at the house for collection? Yes…no…..

2.21 Why will you separate food waste and plastic waste (or NOT separate waste)?………………………………………………………………………

2.22 Which of the following motivate you to sort and separate food and plastic wastes for collection? *(tick one or more)*
☐ Buy your own bins for storing food waste/plastics and pay a reduced user charge
☐ Those NOT separating food/plastic wastes should pay unreduced user charge

□You would like to be provided with free bins
□You would like waste to be collected frequently,
□You would like see clean environment and unchoked drains
□Others...

Payment of user fees for solid waste collection
3.1 It cost Assembly more money for waste collection and disposal. Who should pay for waste collection?
     □Generators only □Generators and Assembly □Government and Assembly
3.2 Do you pay for the collection service? □Yes.........□/no.........
3.3 If yes, how much do you pay per month in Ghana cedis?
     a) For communal collection, GH¢ ......b) For door-to-door collection, GH¢ ...
3.4 If yes, how do you rate the existing tariff? a) High   b) Moderate,   c) Low / affordable
3.5 Would you be willing to pay MORE for the collection service? □YES...... □NO......
3.6 How much are you willing to pay per month in cedis per Household for solid waste collection?
     a) For communal collection, GH¢ ......b) For door-to-door collection, GH¢ ...
3.7 How would you want to be charged for solid waste sorting and separation?
     a.   Pay fixed amount of charge for any amount of waste you generate, and bring for collection
     b.   Pay different amount of charge for different quantities of waste ....
3.8 What is your total household income per month?a.GH¢ <100 b.GH¢101-200 c.GH¢201-400 d.GH¢>401
3.9 Who contributes to household income? ....
3.10 What is your monthly Expenses on:
     a.   Solid waste ....b. toilet.................. c. water......................
     d.   Housing (rent)...............................e. Electricity........................

PRIVATE COMPANIES' QUESTIONNAIRE

General Information
1.1 Date of interview...
1.2 Company's name...
1.3 City and Office Location...Tel No.................................
1.4 Sex of respondent (interviewee)   □ Male □ Female
1.5 Designation or status in the company..
1.6 Indicate your profession..
1.7 Indicate level of terminal formal education □Nil □Basic □Second cycle □Tertiary

2   Technical Operations

2.1 What type of collection service does the company renders? (tick)
☐ Only house-to-house service (curbside/door to door)
☐ Only communal collection service
☐ Both house-to-house service and communal collection services

2.2 Who provides the standard bins for house-to-house service rendered by your company?...

2.3 How many communal containers are owned by the company?..................................

2.4 How many communal containers are picked by the company?..................................

2.5 How many collection vehicles are owned by the company?..................................

2.6 How many collection vehicles are on road?..................................

2.7 How many vehicles are hired by the company?..................................

2.8 Give the collection vehicles types owned by the company, theirs ages and number not on road.

Type of vehicles	Number	Ages of vehicles (years)	Number not on road
Compactor			
Side loader			
Skip truck			
Roll-on-truck			

2.9 Where do you repair your vehicles? ☐ in-house vehicle workshop ☐workshop elsewhere

2.10 If your company has no maintenance workshop, where do you maintain your vehicles? ....

2.11 How many mechanics do you have for in-house repairs?..................................

2.12 How many operations supervisors do you have?..................................

2.13 What is the approximate distance between the service area and the disposal site?
....

Vehicle operations planning

2.14 How would you rate vehicle operations planning?
Scales (1) strongly disagree (2) disagree (3) cannot tell (4) agree (5) strongly agree

a)	Vehicle route plan is prepared for drivers	1	2	3	4	5
b)	Drivers use vehicle route plan for drivers	1	2	3	4	5
c)	Drivers have vehicle log book for recording miles and time	1	2	3	4	5
d)	Drivers use waste pick-up schedules indicating day and time	1	2	3	4	5
e)	Waste pick-up schedules are strictly followed	1	2	3	4	5
f)	Drives use their own discretion about which route to take	1	2	3	4	5

g)	Drivers follow specific routes for solid waste collection	1	2	3	4	5
h)	Records in vehicle log books are analysed by supervisor/engineer	1	2	3	4	5
i)	Traffic studies are conducted and traffic delays avoided	1	2	3	4	5

Who prepare vehicle route plan ....…....

Who prepare vehicle schedules (that is waste pick-up) ....

Vehicle operations Supervision

2.15 Number of operations supervisors... ....

2.16 Number of vehicles supervised by each supervisor ....

2.17 How would you rate vehicle operations supervision?

Scales (1) strongly disagree (2) disagree (3) cannot tell (4) agree (5) strongly agree

a.	Supervisor checks records in vehicle log books	1	2	3	4	5
b.	Supervisor monitors and follow-up waste collection	1	2	3	4	5
c.	Supervisor records the daily number of trips and tonnage of waste	1	2	3	4	5
d.	Management able to manage and analyse log records	1	2	3	4	5
e.	Frequency of waste pick-up are strictly followed	1	2	3	4	5
f.	A scheduled waste pick-up delayed, is picked up within 24 hours	1	2	3	4	5
g.	The drivers and collection crew put on protective clothing	1	2	3	4	5
h.	Waste spillover unto the ground at collection points is collected	1	2	3	4	5
i.	High moral of workers exist due to fair labour practices	1	2	3	4	5

Vehicles Maintenance

2.18 Where do you repair your vehicles? 1. in-house workshop   2. elsewhere

2.19 Number of mechanics for in-house repairs...

2.20 Number of vehicles per mechanic.....................................…….......

2.21 Breakdown vehicle comes back on road within how many days?...........................

2.22 How would you rate vehicle maintenance?

Scales (1) strongly disagree (2) disagree (3) cannot tell (4) agree (5) strongly agree

a)	Vehicle maintenance schedules are prepared and followed	1	2	3	4	5
b)	Breakdowns occur despite planned maintenance	1	2	3	4	5
c)	Minor breakdowns are repaired within 24 hours	1	2	3	4	5
d)	vehicle breakdowns affect waste collection	1	2	3	4	5

2.23 How is vehicle operations supervised?...

2.24 What is your collection frequency per week for house-to-house service?

....

2.25 What is the number of collection crew including driver for each of the following:
    a. Compactor vehicle ....
    b. Skip vehicle, Roll-on Vehicle, Side loader
    c. Tractor...

2.26 What are the reasons for waste spillover at communal container loca-
    tions?...

2.27 How would you judge the cooperation of residents served by house-to-house service?

....

2.28 What complaints are often received from service beneficiaries?............................

2.29 What complaints about service beneficiaries are often received from collection
    crew?...

2.30 How are these complaints responded to?...

Vehicle and traffic conditions

2.31 How would you rate the condition of road used for waste collection
        □ Very poor □ poor □ fair □ good □ very good

2.32 How would you rate the nature of traffic conditions along collection routes
        □ Slow/low congestion □ moderate congestion □ high congestion

2.33 How do 'make' and 'age' of vehicles and the frequency of breakdown for the past
    year affect operations?...

3   Cost Recovery and Financial Aspects

3.1 What is the total number of registered house-to-house users?

    ....

3.2 How is user fee charged for house-to-house service?
        □ Fee per household □ fee per house   □ fee per bin □ fee per volume of waste

3.3 Do you charge the same user fee for all house-to-house service users? □yes □ no

3.4 If NO, indicate the amount and number of customers in the categories below?

No.	Number of households registered in each class	Amount charged per household (¢)	Number of Households who pay regularly
Class 1			
Class 2			
Class 3			
Class 4			

3.5 What is the number of house-to-house users/beneficiaries who pay regularly?

    ....

3.6 How many months are defaulters of fee payment allowed to be in arrears?................................……………………………………………………………

3.7 What sanctions are applied to defaulters of house-to-house service fees payment?................……………………………………………………………………..

3.8 What factors contribute to the defaulting of user fee payment for house-to-house service?................……………………………………………………………..

3.9 What is the total amount billed per month for house-to-house service……..
..........………………………………………………………………………………

3.10 What is the actual total amount collected per month for house-to-house service……..

3.11 What is the cost per tonne of house-to-house waste collected and transported to the disposal site? ……………………………………………….

3.12 How many people are served by communal collection service?..............................

3.13 Do you charge user fee for communal service users? ☐yes ☐ no

3.14 If yes, indicate the amount being paid?...

3.15 If no, why are communal service users not paying user fee?...……………………………........................……

3.16 What is the cost per tonne of communal waste collected and transported to the disposal site? ……………………………………………….

3.17 How are trucks purchases? (Source (s) of funds for purchasing of trucks?...............................……………………

3.18 What is (are) the source(s) for funds for working capital (for paying salaries, buying fuel and repairing vehicle)?..............……………………………….

3.19 What is the return (annual profits) on your investments (in term of percentage of investment)?………………………………….

3.20 How do you judge the return on your investments?..

3.21 How do you judge the financial viability (or profitability) of the waste business?.................

3.22 How do you judge the cost recovery of the waste service?............................
...............................………………………………………........................

3.23 What were the total annual costs of the following items?

items	2004	2005	2006	2007
Fuel consumption cost				
Salaries and benefits				
Vehicle maintenance and servicing				
Depreciation of vehicles				
Withholding tax				
Rent of office space and utilities				
Insurance of collection vehicles				

3.24 What were the total subsidy for house-to-house, annul turnover (total revenue) and total annul expenditure for the past four years?

Year	total subsidy	annul turnover (including subsidy)	total annul expenditure
2004			
2005			
2006			
2007			

4 Company Characteristics and Human Resource Capacity

4.1 What is the type of company □ not registered □limited liability by share □partnership □Sole proprietor □ Limited liability by guarantee (NGO)

4.2 When was the company registered?...

4.3 When did the company commence solid waste services?...........................

4.4 How do you classify the size of the company? □small scale□medium scale □large scale

4.5 How many management levels?..

4.6 Give the order of management control and information flow (example: director→manager→supervisors/mechnics→collection crew/drivers/cleaners)

....

4.7 How are the workers motivated?...

4.8 Has the company sponsored any of the employees to undergo short training to enhance his/her performance?...

4.9 If yes, what kind of training?...

4.10 What kind of study leave do you have for the employees?....................................

4.11 How many people work in the company? Fill the following table.

Posts held	Number of staff	Salary	Highest qualification
Managing Directors *(Part of ownership)*			
Managers/Administrator			
Engineers			
Operations supervisor			
Accountant			
Collection vehicle drivers			
Collection workers (loading of waste)			
Workshop staff (Mechanics)			
Other, specify			

5 Contract and Public-Private Partnership Arrangements

5.1 What type of contract(s) are you operating under □ No contract □franchise contract □Service contract □ other, specify...

5.2 How was the franchise contract procured? □ No bidding □open competition bidding □ other, specify..

5.3 How is the franchise contract renewed?
□ Automatic renewal □competition bidding□ other, specify............................

5.4 How was the communal service contract procured? □ No bidding □open competition bidding □ other, specify..

5.5 How is the communal service contract renewed?
□ Automatic renewal □competition bidding □ other, specify.........................

5.6 How do you judge the transparency and level playing field of the contract processes in solid waste business?...

5.7 Do you think or know of some political interference in awarding contracts to waste contractors ....

5.8 What is the duration of franchise contract?
....

5.9 Is the franchise contract duration enough to recover investment in vehicles?
....

5.10 What is the duration of communal service contract? ....

5.11 Is the communal service contract duration enough to recover investment in vehicles?
....

5.12 Is the house-to-house contract signed with □the Assembly □the residents

5.13 Are the house-to-house service users registered with □the Assembly □the company

5.14 Does the contract have price adjustment formula and cost escalation indices which are tied to the rate of inflation, fuel, labour or consumer price index?... ....

5.15 If there are no price adjustment factors, how are your operations affected?
....

5.16 Is the service area delineated into zones with well known boundaries?
....

5.17 Does your company operate in the zone alone?...

5.18 How do you judge the size of the service area to the capacity of your company?.................. ....

5.19 What is your opinion about the Assembly's obligations under the contract?
Use these Scales (1) strongly disagree (2) disagree (3) cannot tell (4) agree (5) strongly agree

          1 2 3 4 5

(a). Assembly provides designated and accessible landfill site ☐ ☐ ☐ ☐ ☐
(b). Assembly provides designated and accessible communal ☐ ☐ ☐ ☐ ☐
container sites
(c). Assembly provides communal containers (23, 12 and ☐ ☐ ☐ ☐ ☐
10m3)
(d). Assembly provides standard bins for house-to-house ser- ☐ ☐ ☐ ☐ ☐
vice
(e). Assembly promulgates bye-law to assist contractor in the ☐ ☐ ☐ ☐ ☐
service
(f). Assembly conducts public education regularly ☐ ☐ ☐ ☐ ☐
(g). Assembly pays contractor within 45 working days after ☐ ☐ ☐ ☐ ☐
date of claim
(h). Assembly pays interest charges as stipulated in the con- ☐ ☐ ☐ ☐ ☐
tract if payment delayed beyond 3 months

5.20 Is there any political interference in the day-to-day management and operations of
vehicles? …………………………………………………………….........

5.21 Who sets the user fee for house-to-house services? .
………………………………………………………………………...............

5.22 What role do you play in setting user fee for house-to-house services?
……………………………………………………………………….................

5.23 What role do you play in setting user fee for communal services? …………………
……………………………………………………………………….............

5.24 In case you do not participate in setting user fee, do you see the user fee set as reasonable? …………………...……..

5.25 Are the house-to-house services subsidized by Government or Assembly…………..

5.26 Does the Assembly honour its obligation to pay for the service or subsidy on time?
……………………………………………………….

5.27 In the case of delay payment for communal services, how does this affect your operations? …………………...…..

5.28 What is your opinion about honouring of company's obligations under the contract?
*Use these Scales (1) strongly disagree (2) disagree (3) cannot tell (4) agree (5) strongly
agree*

	1	2	3	4	5
(a). Company provides trucks sufficient for solid waste collection	☐	☐	☐	☐	☐
(b). Company achieves daily collection target in the contract	☐	☐	☐	☐	☐
(c). Company disposes off waste at designated landfill	☐	☐	☐	☐	☐
(d). Company uses professional and technical personnel re-	☐	☐	☐	☐	☐

quired
(e). Company keeps complaints of residents/users   ☐ ☐ ☐ ☐ ☐
(f). Company collection crew uses protective clothing   ☐ ☐ ☐ ☐ ☐
(g). Company covers waste containers during transporting   ☐ ☐ ☐ ☐ ☐
(h). Company keeps vehicles in good appearance/sanitary   ☐ ☐ ☐ ☐ ☐
      condition and road worthy,
(i). Company carries out regular repair and maintenance of   ☐ ☐ ☐ ☐ ☐
      containers
(j). Company maintain container site free of litters and clean   ☐ ☐ ☐ ☐ ☐

6   Regulatory (External) Factors Affecting Solid Waste Management

6.1 What performance targets are set in the service con-
tracts?.. ....

6.2 How is the performance targets in the service contract monitored by the Assem-
bly?...……....

6.3 What kind of reports on operations do you submitted to the Assembly?...……………………....…...

6.4 What is the frequency of operations report submitted by your company to the Assembly?.......…………………………………………………………………….................

6.5 What factors outside your control affect your service?...

6.6 How is the quality of service rendered by your company regulated?.........................

6.7 Is the Assembly able to monitor the service contracts?..

6.8 What is the reason why the Assembly is able or not able to monitor the companies'
service quality?...

6.9 What is the Assembly doing to apply the full cost recovery policy?.........….........
....…...…......

6.10 What is the Assembly doing to enforce by-laws?...................…..………...…..

6.11 What is the Assembly doing to enforce national laws?.........................

6.12 Is the Assembly financial independent, and does not request for funds from central
government?.......………………………………………………….........…..……..

6.13 Why is the Assembly not capable or capable of paying your company regularly?
.........…………………………………………………………………….......

6.14 What is the effect of the following on the company's performance?
*Use these Scales (1) high negative effect (2) negative effect (3) no effect (4) positive effect
(5) high positive effect   (........on the performance of company)*

                                                1   2   3   4   5

(a). Frequency of payments of service contract   ☐ ☐ ☐ ☐ ☐
(b). Company's compliance to regulation of service quality   ☐ ☐ ☐ ☐ ☐
(c). Current level of user charges for house-to-house service ☐ ☐ ☐ ☐ ☐
(d).   Current level of charge or price per tonne for service con- ☐ ☐ ☐ ☐ ☐
         tract
(e). Financial autonomy (level of budget) of the Assembly   ☐ ☐ ☐ ☐ ☐
(f).   Autonomy of the Assembly in decision making on the   ☐ ☐ ☐ ☐ ☐
         level of user charge

6.15 How would you judge the implementation of sanitation policies by the Assembly?
*Use these Scales (1) strongly disagree (2) disagree (3) cannot tell (4) agree (5) strongly
agree*

			1	2	3	4	5

(a). City Authority (Assembly) as legal entity has powers con-   ☐ ☐ ☐ ☐ ☐
      ferred on it by the Local Government Act 1993 (Act 462) to
      promulgate by- laws to govern and regulate sanitation.
(b). All solid wastes generated in the city are regularly collected ☐ ☐ ☐ ☐ ☐
      and disposed of in controlled landfills or by other environ-
      mentally acceptable means.
(c). Waste Management Department (WMD) collects at most   ☐ ☐ ☐ ☐ ☐
      20% of the waste collected in the city.
(d). The involvement of the private sector in the waste collection ☐ ☐ ☐ ☐ ☐
      services has improved solid waste management.
(e). City is delineated into zones or service areas and each com-   ☐ ☐ ☐ ☐ ☐
      pany operates in a specific zone.
(f). Each private company has monopoly in a zone.   ☐ ☐ ☐ ☐ ☐
(g). The private sector operates within the policies, regulations, ☐ ☐ ☐ ☐ ☐
      supervisory and licensing arrangements set up by the Assem-
      bly
(h). The policy on full cost recovery where possible (middle and ☐ ☐ ☐ ☐ ☐
      high income areas) is not being pursued seriously.
(i). The by–laws of the Assembly and national laws on environ-   ☐ ☐ ☐ ☐ ☐
      mental sanitation are not enforced
(j). The fine for defaulting by-laws are not punitive (penalizing) ☐ ☐ ☐ ☐ ☐
      enough
(k). The monitoring of compliance to solid waste service stan-   ☐ ☐ ☐ ☐ ☐
      dards and regulations set in the contracts are not done effec-
      tively.
(l). The environmental health standards and sanitary regulations ☐ ☐ ☐ ☐ ☐

are not strictly observed and enforced

6.16 How do you rate the regulation of solid waste services by the Assembly?
Use these Scales (1) very inadequate (2) inadequate (3) normal (4) adequate (5) very
adequate

Assembly's access to information   1   2   3   4   5
(a).   Clear and measurable performance and targets are set in the
       contract
(b).   Assembly's access to information on company's tonnage of   1   2   3   4   5
       waste disposed off
(c).   Assembly's access to information on company cost of waste   1   2   3   4   5
       collection
(d).   The education of the public on environmental and sanitation   1   2   3   4   5
       issues

Regulation of user charges
(e).   Assembly sets unit prices and user charges for house-to-   1   2   3   4   5
       house service alone
(f).   Assembly sets price per tonne for service contract alone   1   2   3   4   5
(g).   Autonomy to set level of user charges   1   2   3   4   5
(h).   Autonomy to set level of price per tonne for service contract   1   2   3   4   5

Regulation of quality of service
(i).   Monitoring of company's service quality and compliance to   1   2   3   4   5
       regulations and standards set the contracts
(j).   Monitoring of quantity of waste collected and outputs of   1   2   3   4   5
       company
(k).   Contract management and supervision of the private sector   1   2   3   4   5
       by the Assembly
(l).   Assembly's adherence to the obligations and regulations in   1   2   3   4   5
       the contracts
(m).   Assembly's enforce sanctions in the contracts signed with   1   2   3   4   5
       companies

Capacity of Assembly
(n).   Accountability of the Assembly to EPA and head Ministry   1   2   3   4   5
(o).   Independence of the Assembly in regulating the company's   1   2   3   4   5
       services
(p).   Resources and capacity of Assembly to monitor service   1   2   3   4   5
       quality and non compliance of service delivery,
(q).   Frequency of monitoring companies' compliance to service   1   2   3   4   5
       quality and standards

6.17 How would you judge the level of autonomy of the Assembly?
Use these Scales (1) very low (2) low (3) normal (4) high (5) high very

		1	2	3	4	5
	Autonomy					
(a).	Autonomy to set level of user charges	☐	☐	☐	☐	☐
(b).	Autonomy to set level of price per tonne for service contract	☐	☐	☐	☐	☐
(c).	Financial autonomy of the Assembly (level of budget for solid waste)	☐	☐	☐	☐	☐
(d).	Autonomy of the Assembly to terminate contract of non-performing company	☐	☐	☐	☐	☐
(e).	Autonomy to enforce by-laws and prosecute free riders	☐	☐	☐	☐	☐
(f).	Autonomy to implement national sanitation policy	☐	☐	☐	☐	☐
(g).	Autonomy of the Assembly in decision making on the level of user charge	☐	☐	☐	☐	☐

6.18 How would you rate the adequacy or otherwise of the following financial issues?
Use these Scales (1) very inadequate (2) inadequate (3) normal (4) adequate (5) very adequate

(a).	Frequency of payments of service contract	1	2	3	4	5
(b).	Current level of user charges for house-to-house service	1	2	3	4	5
(c).	Current level of charge or price per tonne for service con-tract	1	2	3	4	5
(d).	Financial autonomy (level of budget) of the Assembly	1	2	3	4	5

CURRICULUM VITAE

Sampson Oduro-Kwarteng was born on 20th May, 1968 at Dunkwa-on-Offin in Ghana. He enrolled in undergraduate programme in Civil Engineering at the Kwame Nkrumah University of Science and Technology (KNUST), Kumasi, Ghana from 1991 to 1996 and graduated with B.Sc in Civil Engineering. In 1999 he enrolled in M.Sc programme in Water Supply and Environmental Sanitation at KNUST and graduated in 2001. In the last four years, he has been working on his PhD research titled *'Private Sector Involvement in Urban Solid Waste Collection: Performance, capacity and Regulation in five cities in Ghana'*. The PhD was financed by the Dutch Government through a capacity building project titled 'Water Resources and Environmental Sanitation Project' implemented by Kwame Nkrumah University of Science and Technology and UNESCO-IHE Institute for Water Education, Delft, The Netherlands.

He was employed as a lecturer in 2002 at KNUST and since then he has been teaching undergraduate and postgraduate courses in Water Supply, Sanitation, Solid Waste Management and Project/Construction Management in the field of Water and Waste Engineering and Management. He has been involved in research projects and capacity building programmes at the Civil Engineering Department of KNUST in the following areas: water, sanitation and hygiene (WASH), faecal sludge and solid waste management, capacity needs assessment of solid waste providers in Accra (for CHF International), small towns' water supply staff training needs assessment and capacity building, and Kumasi water distribution modelling and management (PoWER). He worked as a Site Engineer on construction projects in Ghana for three years from 1996 to 1999. He has been involving in construction project supervision, consultancy and training including: feasibility studies and engineering designs of community water and sanitation projects. He has ten journal publications and five papers in international refereed conference proceedings.

Contacts: Civil Engineering Department, Kwame Nkrumah University of Science and Technology (KNUST), Kumasi, Ghana. Tel/Fax +233 322 060635, Cell +233 24 4598999. His email address is sokwarteng@yahoo.com.

*For Product Safety Concerns and Information please contact
our EU representative GPSR@taylorandfrancis.com Taylor & Francis
Verlag GmbH, Kaufingerstraße 24, 80331 München, Germany*

T - #0104 - 230425 - C376 - 244/170/20 - PB - 9780415692779 - Gloss Lamination